Introduction
to
# CRYSTALLOGRAPHY
and
# SOLID STATE PHYSICS

# Introduction to CRYSTALLOGRAPHY and SOLID STATE PHYSICS

Debadideb Bhattacharya PhD

Reader and Head (Retd), Department of Physics
Barrackpore Rashtraguru Surendranath College, Barrackpur
Formerly Lecturer in Physics, Ramsaday College, Amta, Howrah

New Central Book Agency (P) Ltd

LONDON

HYDERABAD ERNAKULAM BHUBANESWAR

NEW DELHI KOLKATA PUNE GUWAHATI

**NCBA**

REGD OFFICE
8/1 Chintamoni Das Lane, Kolkata 700 009, India
email: ncbapvtltd@eth.net

OVERSEAS
NCBA (UK) Ltd, 149 Park Avenue North, Northampton, NN3 2HY, UK
email: ncbauk@yahoo.com

EXPORT
NCBA Exports Pvt. Ltd, 212 Shahpur Jat, New Delhi 110 049
email: ncbaexports@ncbaindia.com

BRANCHES

208B Peacock Lane, Shahpur Jat
**New Delhi** 110 049
email: ncbadel@ncbapvtltd.com

House No. 3-1-315, 1st Floor
Nimboliadda, Kachiguda, **Hyderabad** 500 027
email: ncbahydb@ncbapvtltd.com

Shop Nos. 3 & 4, Vinayak Towers
681/B Budhwar Peth, Appa Balwant Chowk
**Pune** 411 002
email: ncbapune@ncbapvtltd.com

GSS Shopping Complex, 1st Floor
Opposite College Post Office, Convent Road
**Ernakulam** 682 035
email: ncbaernk@ncbapvtltd.com

Shop No. 15, Bharati Towers, Ground Floor
Forest Park, **Bhubaneswar** 751 001
email: ncbabub@ncbapvtltd.com

Radhamayee Complex (Second Floor)
Opp. SBI, Panbazar, Jaswanta Road
**Guwahati** 781 001
email: ncbaguwahati@ncbapvtltd.com

INTRODUCTION TO CRYSTALLOGRAPHY AND SOLID STATE PHYSICS • Debadideb Bhattacharya



First Published: August 2010
Thoroughly Revised and Updated Second Edition: July 2013

PUBLISHER AND TYPESETTER
New Central Book Agency (P) Ltd
8/1 Chintamoni Das Lane, Kolkata 700 009

PRINTER
New Central Book Agency (P) Ltd
Web-Offset Division, Dhulagarh, Sankrail, Howrah

COVER DESIGNER
Soumen Paul

COVER PRINTER
Biswanath Studio, Kolkata

TECHNICAL EDITOR
Dipan Roy

PROJECT TEAM
Prabhat Jas, Prodip Baidya, Pradip Biswas, and Swapan Sarkar

ISBN 978 81 7381 631 4

Price ₹ 415.00

# Contents

**Preface to the Second Edition** ix
**Preface to the First Edition** x

**1. Crystal Structure and X-ray Diffraction** 1–38

1.0 Introduction 1
1.1 Crystalline and Amorphous Solids 1
1.2 Crystal Translation Vectors and Lattices 2
1.3 Unit Cells and Primitive Cells 3
1.4 Crystal System 4
1.5 Crystal Structure 7
1.6 Relation between the Density of Crystalline Material and Lattice Constant in a Cubic Lattice 13
1.7 Miller Indices 14
1.8 Spacing of Planes in Crystal Lattice 15
1.9 Reciprocal Lattice 18
1.10 X-rays 19
1.11 Origin of X-rays 19
1.12 X-ray Spectrum 20
(i) Continuous X-ray spectrum 20
(ii) Characteristic X-ray spectrum 21
1.13 Mosley's Law 22
1.14 Absorption of X-rays 23
1.15 X-ray Diffraction: Laue's and Bragg's Equations 24

1.16 X-ray Diffraction and Bragg's Law of Crystal 25
1.17 Experimental Method of X-ray Diffraction 26
**Solved Problems** **28**
**Review Questions** **36**
**Problems** **37**

**2. Bonding in Solids** **39–43**

2.0 Introduction 39
2.1 Molecular Crystals 40
2.2 Ionic Crystals 41
2.3 Metallic Crystals 42
2.4 Covalent Crystals 42
2.5 Hydrogen Bonded Crystals 42
2.6 Differences between Ionic, Covalent, and Metallic Bonding in Solids 43
**Review Questions** **43**

**3. Free Electron Theory of Solids** **44–57**

3.0 Introduction 44
3.1 Free Electron Model 44
3.2 Deduction of Wiedemann-Franz Law 45
3.3 Sommerfeld's Model or Quantum Theory of Free Electrons 46
3.4 Free Electron Theory of Sommerfeld in Three Dimensions 50
**Solved Problems** **55**
**Review Questions** **57**
**Problems** **57**

**4. Band Theory of Solids** **58–77**

4.0 Introduction 58
4.1 The Bloch Theorem 59
4.2 Bloch Theorem in Three Dimensions 62
4.3 The Kronig-Penney Model 63
4.4 The Velocity of the Electron according to the Band Theory 68
4.5 The Distinctions between Metals, Insulators, and Intrinsic Semiconductors 71
4.6 Drift Velocity, Mobility, and Conductivity of Metal 73
4.7 Hall Effect in Metal 73
**Solved Problems** **74**
**Review Questions** **76**
**Problems** **76**

**5. Magnetic Properties of Material** **78–95**

5.0 Introduction 78
5.1 Types of Magnetism 79
5.2 Langevin's Theory of Diamagnetism (Classical Theory) 79
5.3 Langevin's Theory of Paramagnetism (Classical Theory) 82
5.4 Quantum Theory of Paramagnetism 84

5.5 Ferromagnetism 88
5.6 Differences among Ferromagnetic, Paramagnetic, and Diamagnetic Substances 93
**Solved Problems** **93**
**Review Questions** **94**
**Problems** **95**

**6. Semiconductor** **96–111**

6.0 Introduction 96
6.1 Intrinsic Semiconductor 96
6.2 Extrinsic or Impurity Semiconductor 98
6.3 Drift Velocity, Electrical Conductivity, and Carrier Mobility 100
6.4 Carrier Concentration and Fermi Levels in Intrinsic Semiconductor 101
6.5 Model for an Impurity Semiconductor 105
**Solved Problems** **109**
**Review Questions** **110**
**Problems** **110**

**7. Defects in Crystals** **112–114**

7.0 Introduction 112
7.1 Point Defects 113
7.2 Dislocation 113
7.3 Planar Defects 114
**Review Questions** **114**

**8. Superconductivity** **115–128**

8.0 Introduction 115
8.1 Effect of Magnetic Field on Superconductor 116
8.2 Meissner Effect 116
8.3 Specific Heat 118
8.4 London Equation 119
8.5 London Penetration Depth 120
8.6 Type I and Type II Superconductors 122
8.7 BCS Theory of Superconductivity 123
8.8 Tunnelling 123
8.9 Josephson Effects 124
8.10 Applications of Superconductor 125
**Solved Problems** **126**
**Review Questions** **127**
**Problems** **128**

**9. Dielectric Properties of Solids** **129–140**

9.0 Introduction 129
9.1 Electric Polarization 129
9.2 Local Field 131
9.3 Clausius-Mossotti Relation 131

9.4 Types of Polarizability 133
9.5 Dielectric Relaxation and Loss 138
9.6 Ferroelectric Crystal 138
9.7 Piezoelectricity 138
**Solved Problems** 139
**Review Questions** 140
**Problems** 140

**10. Short Biographies of Noted Solid State Physicists** **141–146**
10.0 Introduction 141
10.1 Hendrik Antoon Lorentz 141
10.2 William Henry Bragg 142
10.3 Pierre Ernest Weiss 142
10.4 Arnold Sommerfeld 142
10.5 Paul Langevin 143
10.6 Albert Einstein 143
10.7 Max von Laue 144
10.8 William Lawrence Bragg 144
10.9 Satyendranath Bose 145
10.10 Wolfgang Ernst Pauli 145
10.11 Walter Honser Brattain 146
10.12 Felix Bloch 146
10.13 John Bardeen 146
10.14 William Bradford Shockley 146

**Appendix** **147–153**
**Bibliography** **155**
**Author Index** **157**
**Subject Index** **159–161**

# *Preface to the Second Edition*

In this second edition of *Introduction to Crystallography and Solid State Physics* the author ackowledges with deep sense of gratitude the overwhelming response received from the students and teachers throughout the country. In this edition this book has been significantly edited and enlarged. A good number of new questions and problems have been incorporated to facilitate better and deeper understanding of the multifarious topics.

The author is sincerely thankful to Prof. Sankha Das of Serampore College for his invaluable suggestions for the improvement of this edition. I am also thankful to Sri Amitabha Sen, Director, New Central Book Agency (P) Ltd, and Senior Manager Sri R. M. Nath and his colleagues for their kind interest in the production of this book. The author is also thankful to Sri Dipan Roy, Technical Editor, for his sincere efforts to improve the book.

The author's efforts will be amply rewarded if the students are benefitted from this book. I shall gratefully acknowledge constructive suggestions and criticisms about the book.

21 May 2013 **Debadideb Bhattacharya**
Serampore

# *Preface to the First Edition*

The author feels that there is no single book on Crystallography and Solid State Physics which covers fully and comprehensively the syllabus of the Indian universities though there is a large number of reputed books by Indian and foreign authors on the subject. This book is written with the intention to satisfy fully the multi-dimensional requirements of the Degree students of different Indian universities.

The subject matter has been developed in such a way that the students will be able to progressively sail through the different chapters with confidence and ease. The text is divided into ten chapters with solved problems, review questions, and unsolved problems. The books and journals consulted have been referred to at the end of the treatise.

The author thanks Sri P. K. Roychowdhuri for going through the manuscript and Sri Debapriya Bhattacharya, my son, for giving necessary and constructive suggestions to improve the book. I am also thankful to Sri Amitabha Sen, Director, New Central Book Agnecy (P) Ltd, and Senior Manager, Sri R. M. Nath, and his production colleagues, for their proactive interest in the book.

Author's effort will be amply rewarded if the students are benefitted from this book. I shall gratefully acknowledge any constructive suggestions and criticisms on the book.

1 May 2010 **Debadideb Bhattacharya**
Serampore

# 1

# *Crystal Structure and X-ray Diffraction*

## 1.0 INTRODUCTION

The name *crystal* comes from the Greek word 'Krustallos' meaning 'clear ice'. But later in the seventeenth century the name crystal was extended to manifestation of solids. All materials are composed of atoms and molecules. Atoms are arranged in regular ordered repetitive manner in crystals. We should consider the nature of arrangement of atoms and molecules within the crystalline material. The atoms are in constant motion in actual crystal. The basic feature of all crystals is its regularity in atomic arrangement. Main feature of the crystal structure is the periodicity of their pattern. Crystals are a three-dimensional array of atoms. Crystal also possesses symmetry property. Crystal structures are the basic need of the Solid State Physics. All metals form crystalline structure under normal condition of cooling from molten state. The study on the geometrical form and other properties of crystalline solids by using X-rays, electron beam, γ-rays constitutes the science of crystallography. So to study the Solid State Physics we go to crystallography first.

## 1.1 CRYSTALLINE AND AMORPHOUS SOLIDS

A solid having regularity in atomic arrangement in the form of periodic array is termed as a crystal. Most solid substances when pure are obtainable in crystalline form. The most crystal structures possess the property of symmetry. The crystal generally is of two types—single crystal and

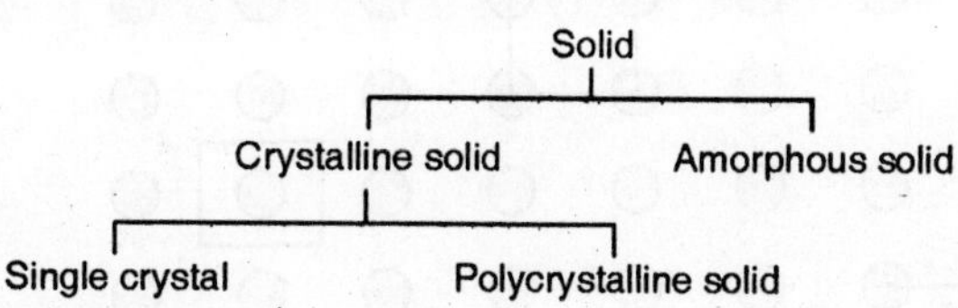

polycrystalline solid. Substances which are without any regular form or symmetrical shape are known as amorphous or in other words, solids which do not form crystals are said to be amorphous. The crystalline solids have directional properties, i.e., the physical properties like thermal conductivity, electrical conductivity have different values in different directions and, therefore, they are called anisotropic substances. Examples of metallic crystals are copper, aluminium, silver, and non-metallic crystals are crystalline carbon, diamond, crystal polymer. The amorphous solids have no directional properties. Examples of amphorphous solids are glasses, plastics etc. The structure of crystalline solids are determined by X-ray diffraction, electron diffraction, γ-rays scattering.

An arrangement of an infinite array of points in the form of a two-dimensional or three-dimensional network in a crystal is called *Space lattice.*

## 1.2 CRYSTAL TRANSLATION VECTORS AND LATTICES

Atomic arrangement in ideal crystal is a net-like arrangement known as *lattice*. A lattice is a regular periodic arrangement of points in space. It is also a mathematical abstractor of points in space. Lattice points denote the position of the atoms in the crystal.

A crystal structure is formed by associating with every lattice point in a unit assembly known as *basis.* A basis is an assembly of atoms identical in composition arrangement and orientation. The logical relation is

lattice + basis = crystal structure.

An ideal crystal is composed of atoms arranged on a lattice defined by three translation vectors $\vec{a}$ , $\vec{b}$ and $\vec{c}$ such that all the lattice points are connected by translation operator $\vec{T}$ which is given by

$$\vec{T} = n_1\vec{a} + n_2\vec{b} + n_3\vec{c} \,, \tag{1.1}$$

where $n_1$, $n_2$, $n_3$ are known as *translation group of crystals.* Two-dimensional arrangements array of points are shown in Figure 1.1.

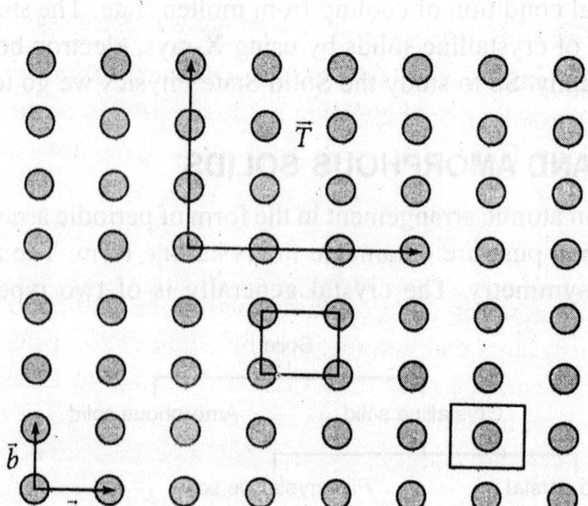


**Figure 1.1** Two-dimensional array of points

Geometrical arrangement of atoms is a function $f(r)$ which describes any property of crystal to be periodic, i.e.,

$$f(\vec{r}) = f(\vec{r} + \vec{T}) \tag{1.2}$$

In general, the atomic arrangement looks the same in every respect when viewed from any point and from another point $\vec{r}$. In general,

$$\vec{r}' = \vec{r} + n_1\vec{a} + n_2\vec{b} + n_3\vec{c} \tag{1.3}$$

## 1.3 UNIT CELLS AND PRIMITIVE CELLS

### (a) Unit cells

The fundamental feature of a crystal is the periodicity in structure. It is convenient to subdivide the crystal structure into small repeat entities called *unit cells,* i.e., in every crystal some fundamental grouping is repeated. Such fundamental grouping of particles is called a unit cell. A group of atoms of definite shape and size is also known as a unit cell. It is observed that a parallelogram such as EFGH is repeatedly translated by the vectors $\vec{a}$ and $\vec{b}$ respectively corresponding to EF and EH. Thus, the whole pattern may be obtained, and EFGH is called a *unit cell* (Figure 1.2).

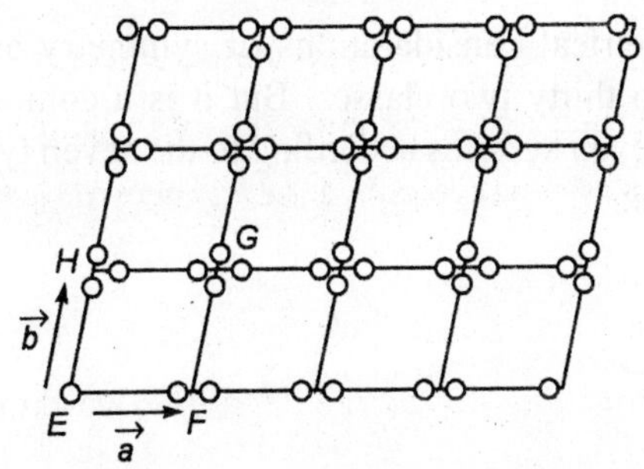


**Figure 1.2** Two-dimensional crystal and various unit cells

If we consider a space lattice which is a part of a three-dimensional crystal, then a unit cell is defined as that volume of the solid from which the entire crystal can be constructed by translation repetition in three dimensions. The true unit cell corresponds to the three basic translation vectors $\vec{a}$, $\vec{b}$ and $\vec{c}$.

### (b) Primitive cells

Primitive unit cell is the smallest unit cell in volume that can be difined for a given lattice and the set of linearly independent vectors that define a primitive unit cell as primitive basis vectors. The unit cell differs from the primitive cell in that it is not restricted to being the equivalent of one lattice point. Unit cells may be primitive cells but all the primitive cells need not be unit cells. The three translation vectors selected to edges of the unit cell in general are noncoplanar and are called *crystallographic axes* $\vec{a}$, $\vec{b}$, $\vec{c}$, and angles between them $\alpha$, $\beta$, $\gamma$ in Figure 1.3, where $\alpha$ is the angle between $\vec{b}$ and $\vec{c}$, $\beta$ is the angle between $\vec{c}$ and $\vec{a}$ and $\gamma$ is the angle between $\vec{a}$ and $\vec{b}$. The angles $\alpha$, $\beta$ and $\gamma$ are called *interfacid angles.* Fundamental translation vectors $\vec{a}$, $\vec{b}$ and $\vec{c}$

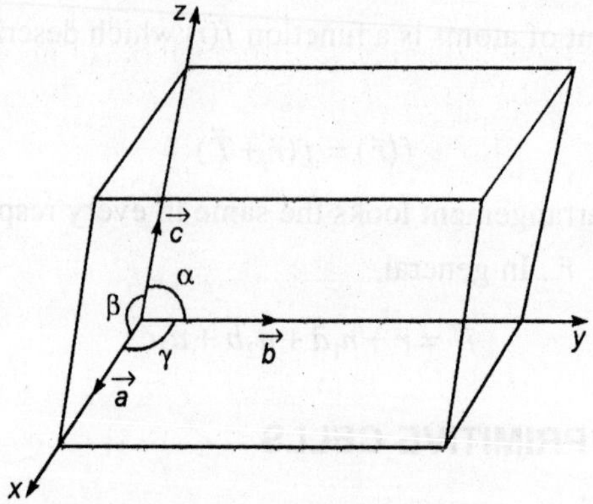


Figure 1.3 Crystal axes and basis lattice parameters of a unit cell

denote the size of unit cell and is known as *primitives*. If the cell contains additional one or more atoms with an atom at each corner of the unit cell, these cells are called *non-primitive cells*.

## 1.4 CRYSTAL SYSTEM

Bravais first introduced the concept of the three-dimensional lattice while explaining crystal structure.

According to the geometrical considerations of symmetry and internal structure, crystal systems can be classified into thirty two classes. But it is a common practice to divide all the crystals having grouped into seven systems according to the seven types of conventional unit cell. The basic crysrtal systems are:

(i) Triclinic
(ii) Monoclinic
(iii) Orthorhombic
(iv) Tetragonal
(v) Cubic
(vi) Trigonal
(vii) Hexagonal.

**(i) Triclinic crystal system**

In the triclinic system the single lattice type has a primitive unit cell with three axes which are not perpendicular to each other and are of unequal lengths and unequal angles. Example: $CuSO_4$, $5H_2O$; $K_2SO_4$ crystal.

$$a \neq b \neq c, \quad \alpha \neq \beta \neq \gamma \neq 90°.$$

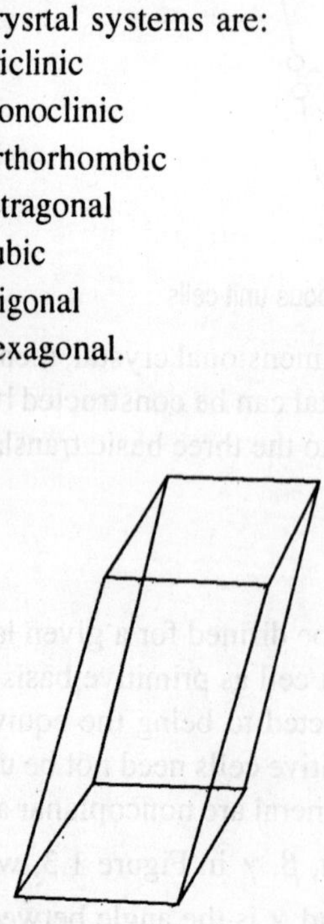

Figure 1.4 Trichlinic crystal

### (ii) Monoclinic crystal system

Two of the crystal axes are perpendicular to each other but the third one is obliquely inclined, and crystal axes are of unequal lengths. Example: $FeSO_4$ , $Na_2SO_4$ crystal, $Na_2SO_4$.

$$a \neq b \neq c, \quad \alpha = \beta = 90° \neq \gamma.$$

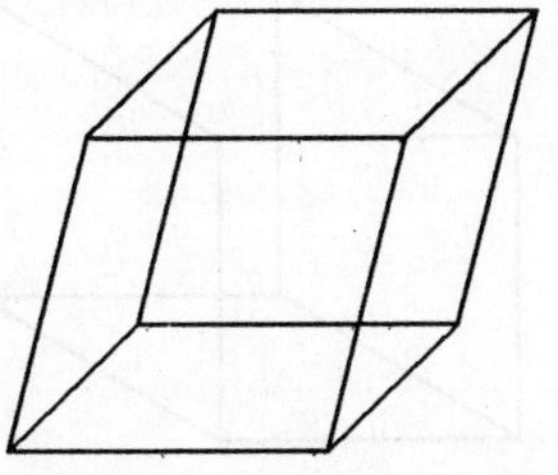

**Figure 1.5** Monoclinic crystal

### (iii) Orthorhombic crystal system

In orthorhombic crystal system, axes are perpendicular to each other and are of unequal lengths. Orthorhombic lattices may be simple, base-centred, body-centred and face-centred. Example: $MgSO_4$, $KNO_3$ crystal.

$$a \neq b \neq c, \quad \alpha = \beta = \gamma = 90°.$$

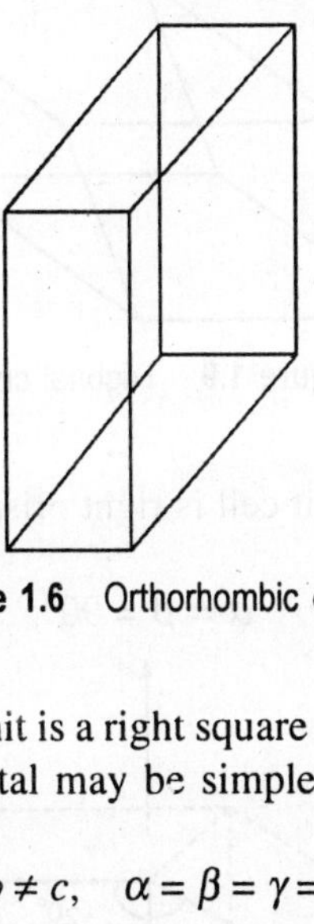

**Figure 1.6** Orthorhombic crystal

### (iv) Tetragonal crystal system

In the tetragonal system, the simplest unit is a right square prism. The crystal axes are perpendicular to one another. The tetragonal crystal may be simple or body-centred crystal. Example: Sn crystal, $TiO_3$.

$$a = b \neq c, \quad \alpha = \beta = \gamma = 90°.$$

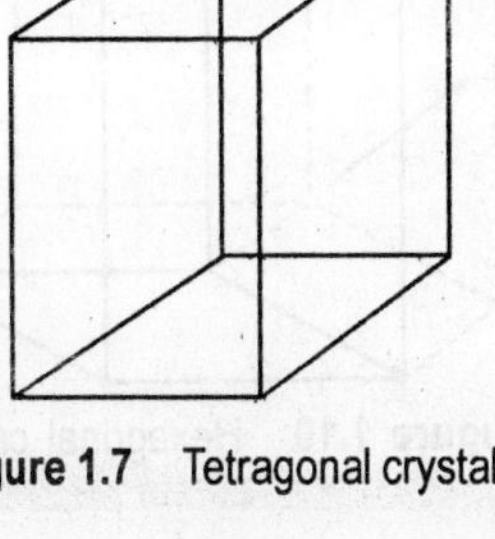

**Figure 1.7** Tetragonal crystal

**(v) Cubic crystal system**

In the cubic crystal system, axes are perpendicular to each other and are of equal lengths. Cubic lattices may be simple, face-centred or body-centred. Example: NaCl crystal, Cu, Al.

$$a = b = c, \quad \alpha = \beta = \gamma = 90°.$$

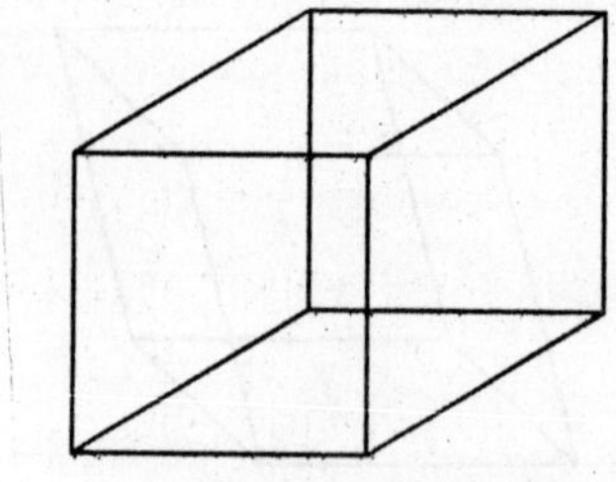

**Figure 1.8** Cubic crystal

**(vi) Trigonal crystal system**

In the trigonal crystal system, rhombohedron is usually chosen as the unit cell. So the axes are equal in length and inclined to each other at 90°. Example: Sb crystal, $CaSO_4$.

$$a = b = c, \quad \alpha = \beta = \gamma \neq 90°.$$

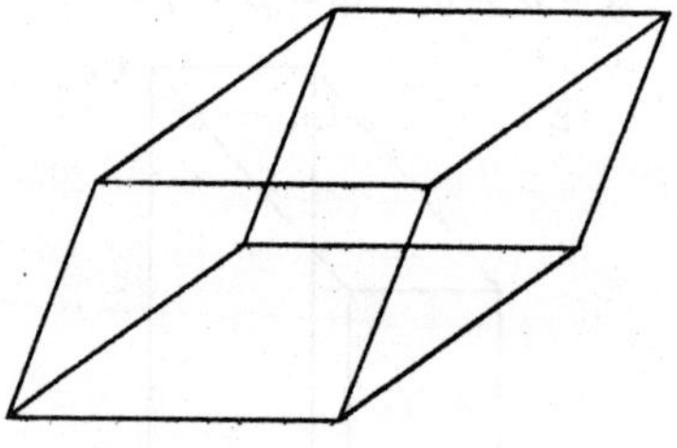

**Figure 1.9** Trigonal crystal

**(vii) Hexagonal crystal system**

In the hexagonal crystal system the unit cell is right prism based on a rhombus with an angle of 120°. Example: AgCl crystal, Mg, Zn.

$$a = b \neq c, \quad \alpha = \beta = 90°, \quad \gamma = 120°.$$

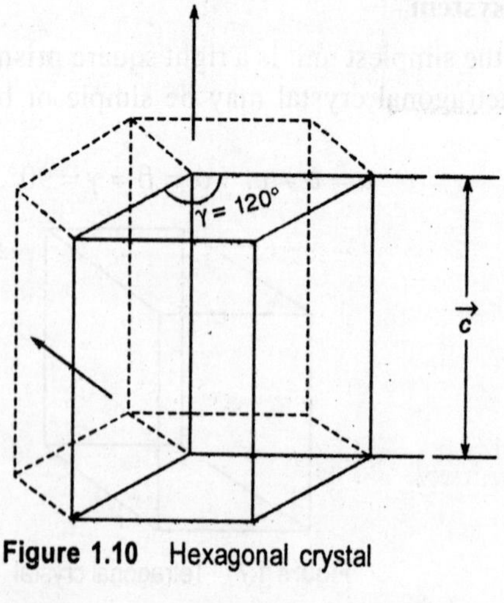


**Figure 1.10** Hexagonal crystal

## 1.5 CRYSTAL STRUCTURE

We discuss relatively simple crystal structure, viz., cubic, body-centred and face-centred crystal structures. To a close approximation we may record each atom as a sphere of definite radius, and crystals are considered to be formed by such spheres and are packed in most stable manner. Some important crystal structure terms are defined below:

*Atomic radius*: Atomic radius ($r$) is defined as half the distance between the nearest neighbours in a crystal of pure element. The distance between the centre of any two neighbour atoms is called *nearest neighbour distance*. The nearest neighbour distance is $2r$ when atomic radius is $r$.

*Coordination number* ($N$): The coordination number ($N$) in a crystal is defined as the number of equivalent nearest neighbours to an atom in a given structure. The closeness of packing of the atoms in the structure gives greater the coordination number.

*Atomic packing factor*: Atomic packing factor or atomic packing fraction is the ratio of the volume occupied by atoms ($v$) to the volume of the unit cell ($V$) relating to the structure of the crystal.

$$\text{So atomic packing factor (APF)} = \frac{v}{V}.$$

(i) *Simple cubic structure:* The simple and easiest structure is the simple cubic structure. The unit cube of this structure has atoms only at the eight corners of the unit cell. This structure is loosely packed and each atom is surrounded by six equivalent nearest neighbours, hence coordinate number of a simple cubic lattice is six. In simple cubic cell, each corner atom is shared by a total number of eight unit cells, hence the share of each corner atom to a unit cell is 1/8 of an atom. Hence, the total number of atoms per unit cell is $1/8 \times 8 = 1$ atom. In a simple cubic structure the atoms touch each other along the edges, hence the atomic radius is given by half the primitive cell, i.e., $r = a/2$, (Figure 1.12).

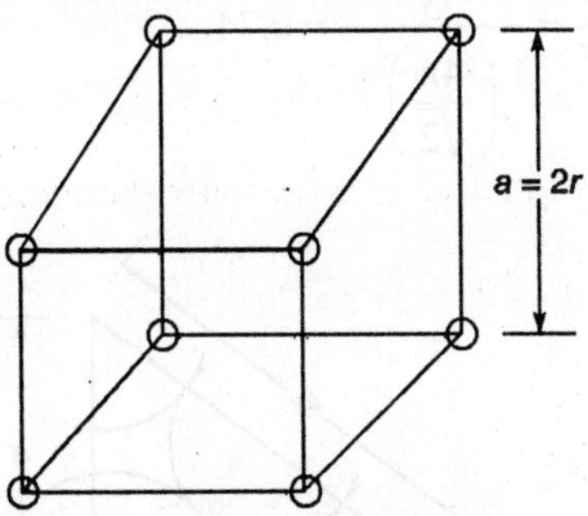


**Figure 1.11** Simple cubic structure

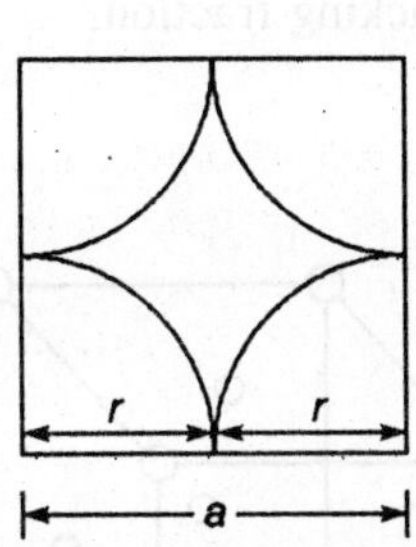


**Figure 1.12** Simple cubic structure

Volume occupied by the atom in a unit cell is

$$v = \frac{4}{3}\pi r^3 = \frac{4}{3}\pi\left(\frac{a}{2}\right)^3.$$

Volume of a unit cell $V = a^3$.

$$\text{Packing fraction} = \frac{v}{V} = \frac{\frac{4}{3}\pi\left(\frac{a}{2}\right)^3}{a^3} = 0.524$$

Polonium at a certain temperature and Potassium chloride show simple cubic structure.

(ii) *Face-centred cubic structure* (FCC): In the face-centred cubic structure, there are atoms at the eight corners of a unit cell and six face-centred atoms. Each corner atom is shared by eight surrounding unit cells and each face-centred atom is shared by two unit cells.

The total number of atoms per unit face-centred structure is $\frac{1}{8}\times 8+\frac{6}{2}=4$. Thus, in a FCC structure there are four atoms per unit cell.

Coordination number = 12.

Now $$(4r)^2 = 2a^2$$

∴ atomic radius, $$r = \frac{\sqrt{2}}{4}a$$

Volume of all atoms in the unit cell, $v = 4 \times \frac{4}{3}\pi r^3$

Volume of the unit cell, $V = a^3 = \left(\frac{4r}{\sqrt{2}}\right)^3$

Atomic packing fraction, $$\frac{v}{V} = \frac{4\times\frac{4}{3}\pi r^3}{\left(\frac{4r}{\sqrt{2}}\right)^3} = 0.74.$$

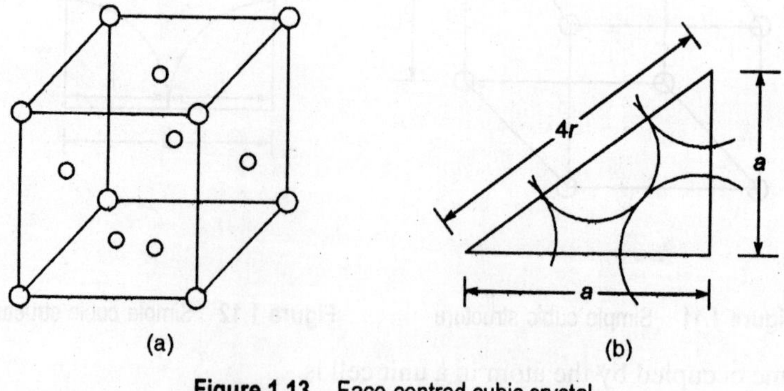


**Figure 1.13** Face-centred cubic crystal

Copper, Nickel, Gold, Silver are the common examples of FCC structure.

(iii) *Body-centred cubic structure* (BCC): In this structure the unit cell is cubic with an atom in the centre of the cube. Number of atoms per unit body-centered cubic cell is $1/8 \times 8 + 1 = 2$.

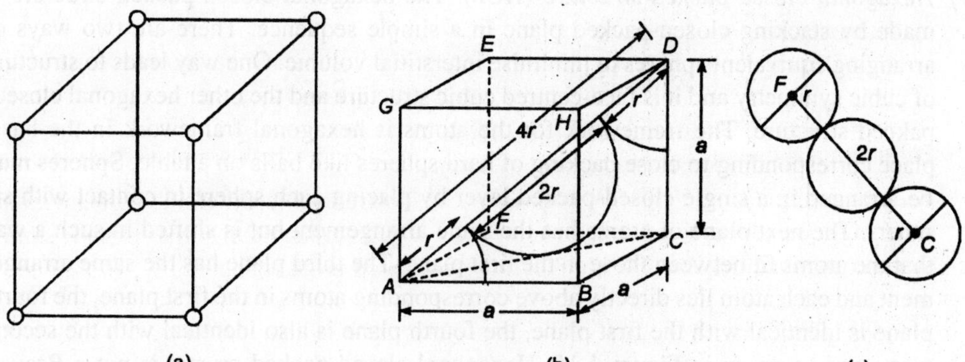


**Figure 1.14** Body-centred cubic structure

Now in Figure 1.14 (b), $AC^2 = AB^2 + BC^2$

$$= a^2 + a^2$$

$$= 2a^2.$$

$$\therefore \quad AC = \sqrt{2}a \cdot$$

Again, $AD^2 = AC^2 + CD^2$

$$= 2a^2 + a^2$$

$$= 3a^2$$

or, $(4r)^2 = 3a^2$

or, $a = \dfrac{4}{\sqrt{3}} r .$

Volume of the cube, $V = a^3 = \left(\dfrac{4}{\sqrt{3}} r\right)^3$

Volume of all atoms in a unit cell, $v = 2 \times \dfrac{4}{3} \pi r^3 .$

$\therefore$ atomic packing fraction, $\dfrac{v}{V} = \dfrac{2 \times \frac{4}{3} \pi r^3}{a^3}$

$$= \frac{2 \times \frac{4}{3} \pi r^3}{\left(\frac{4}{\sqrt{3}} r\right)^3} = \frac{\sqrt{3}\pi}{8}$$

$$= 0.68.$$

Body-centred cubic sturcture (BCC) is not so closely packed as face-centred cubic structure (FCC). Li, Na, K, Cr exhibit BCC structures.

(iv) *Hexagonal closed-packed structure* (HCP): The hexagonal closed-packed structure is made by stacking closed-packed plane in a simple sequence. There are two ways of arranging equivalent spheres to minimise interstitial volume. One way leads to structure of cubic symmetry and it is face-centred cubic structure and the other hexagonal closed-pakced structure. The framework for the atoms is hexagonal framework in the basal place corresponding to close packing of hard spheres like balls on a table. Spheres may be arranged in a single closed-packed layer by placing each sphere in contact with six others. The next plane of atoms has the same arrangement but is shifted in such a way that the atoms fit between those in the first plane. The third plane has the same arrangement and each atom lies directly above corresponding atoms in the first plane, the fourth plane is identical with the first plane, the fourth plane is also identical with the second plane and so on as in Figure 1.15. Hexagonal closed-packed crystal is not a Bravais lattice. The $c/a$ ratio for hexagonal packing of sphere is $\left(\frac{8}{3}\right)^{1/2} = 1.633$, but for most of the materials having this structure the actual ($c/a$) ratio departs from ideal value. Packing fraction in the ideal case of closest packing is $r = a/2$.

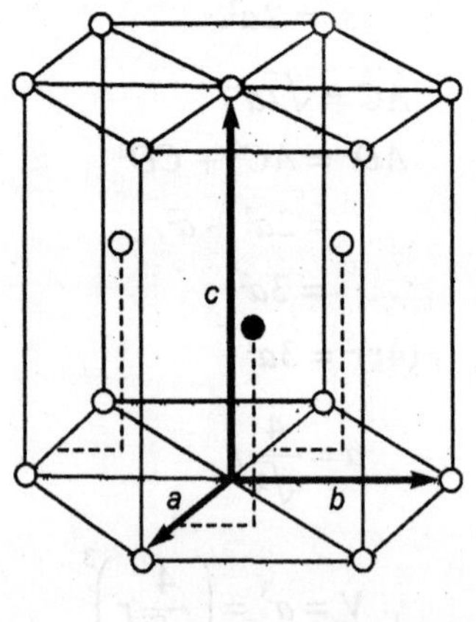


**Figure 1.15** Hexagonal closed-packed structure

$$\text{Packing fraction} = \frac{\text{volume occupied by atoms}}{\text{volume of the unit cell}}$$

$$= \frac{4 \times \frac{4}{3}\pi r^3}{\sqrt{3}\, a^3 c}$$

$$= 0.74.$$

Cd, Mg, Zn are the common examples of hexagonal closed-packed structure.

(v) *Sodium chloride structure*: Sodium chloride is a combination structure of sodium and chlorine. NaCl structure is shown in Figure 1.16. The Bravais lattice is face-centred cube, the basis consists of one Na atom and one Cl atom separated by half the diagonal of a unit cube.

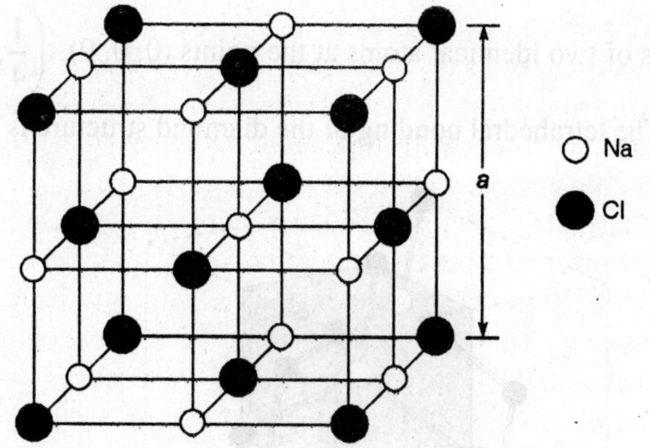


Figure 1.16 Structure of sodium chloride crystal

It can be considered as two face-centred cube sub-lattices of Na ions having its origin at the point (0, 0, 0), the Cl ions having its origin midway along a cube edge, say, at the point ($a$/2, 0, 0). There are four units of NaCl in each unit cube with atoms in positions.

$$\text{Na }(0,0,0), \left(\frac{1}{2}, \frac{1}{2}, 0\right), \left(\frac{1}{2}, 0, \frac{1}{2}\right), \left(0, \frac{1}{2}, \frac{1}{2}\right)$$

$$\text{Cl }\left(\frac{1}{2}, \frac{1}{2}, \frac{1}{2}\right), \left(0, 0, \frac{1}{2}\right), \left(0, \frac{1}{2}, 0\right), \left(\frac{1}{2}, 0, 0\right)$$

Each atom has six atoms of opposite kind as nearest neighbour. Crystals having similar structure like NaCl structure are KCl, KBr, MnO, AgBr, etc.

(vi) *Diamond structure*: The diamond structure is a combination of two inter-penetrating face-centered cubic sub-lattices and is formed by carbon atoms as shown in Figure 1.17.

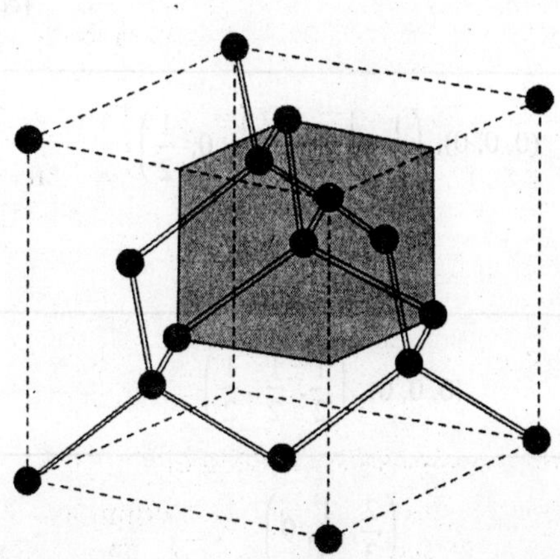

Figure 1.17 Crystal structure of diamond

The primitive basis of two identical atoms at the points (0, 0, 0), $\left(\frac{1}{4}, \frac{1}{4}, \frac{1}{4}\right)$ is associated with each lattice point. The tetrahedral bonding of the diamond structure is shown in Figure 1.18.

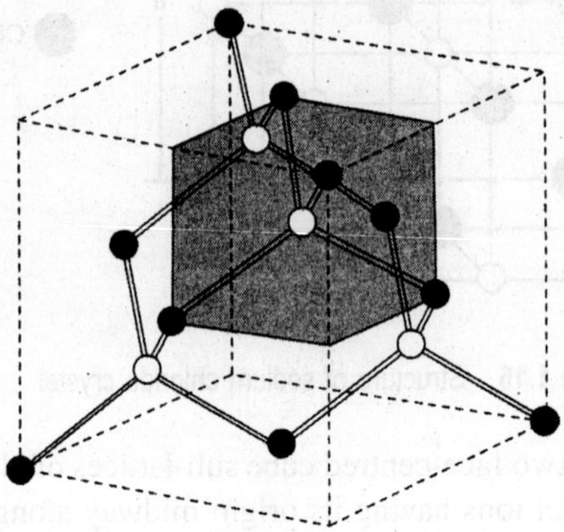

**Figure 1.18** Crystal structure of cubic zinc sulphide

There are eight atoms in a unit cube. The diamond cubic structure is loosely packed, since each atom has four nearest neighbours. The coordination number of the diamond lattice is four. It is not a Bravais lattice. Carbon, silicon, germanium and grey tin crystallise with diamond structure. The diamond structure is a result of covalent bonding.

In Table 1.1 are given the essential properties of several crystal structures.

**Table 1.1**

| *Structure type* | *Number of atoms per unit cell* | *Position of atoms in unit cell (basis coordinates)* | *Number of nearest neighbours (coordination number)* | *Distance of nearest neighbours* |
|---|---|---|---|---|
| Face-centred cubic structure (Cu type) | 4 | (0, 0, 0), $\left(\frac{1}{2}, \frac{1}{2}, 0\right)$, $\left(\frac{1}{2}, 0, \frac{1}{2}\right)$, $\left(0, \frac{1}{2}, \frac{1}{2}\right)$ | 12 | $\frac{a}{\sqrt{2}}$ |
| Body-centred cubic structure | 2 | (0, 0, 0), $\left(\frac{1}{2}, \frac{1}{2}, \frac{1}{2}\right)$ | 8 | $\frac{\sqrt{3}}{2}a$ |
| Hexagonal closed-packed structure | 2 | $\left(\frac{2}{3}, \frac{1}{3}, 0\right)$ | 12 | $\left(\frac{a^2}{3}+\frac{c^2}{4}\right)^{1/2}$ |

| *Structure type* | *Number of atoms per unit cell* | *Position of atoms in unit cell (basis coordinates)* | *Number of nearest neighbours (coordination number)* | *Distance of nearest neighbours* |
|---|---|---|---|---|
| NaCl structure | (Na) 4<br>(Cl) 4 | Na $(0, 0, 0)\left(\frac{1}{2}, \frac{1}{2}, 0\right)$, $\left(\frac{1}{2}, 0, \frac{1}{2}\right)\left(\frac{1}{2}, 0, 0\right)$,<br>Cl $\left(\frac{1}{2}, \frac{1}{2}, \frac{1}{2}\right), \left(0, 0, \frac{1}{2}\right)$, $\left(0, \frac{1}{2}, 0\right), \left(\frac{1}{2}, 0, 0\right)$ | 6 | $\frac{a}{2}$ |
| Diamond structure | 8 | $(0, 0, 0), \left(\frac{1}{2}, \frac{1}{2}, 0\right), \left(\frac{1}{2}, 0, \frac{1}{2}\right)$, $\left(0, \frac{1}{2}, \frac{1}{2}\right), \left(\frac{1}{4}, \frac{1}{4}, \frac{1}{4}\right), \left(\frac{3}{4}, \frac{3}{4}, \frac{1}{4}\right)$, $\left(\frac{3}{4}, \frac{1}{4}, \frac{3}{4}\right) \left(\frac{1}{4}, \frac{3}{4}, \frac{3}{4}\right)$ | 4 | $\frac{\sqrt{3}}{4}a$ |

## 1.6 RELATION BETWEEN THE DENSITY OF CRYSTALLINE MATERIAL AND LATTICE CONSTANT IN A CUBIC LATTICE

Let us consider a cubic crystal with lattice constant $a$, density of crystalline material $\rho$ and atomic weight $M$. If $P$ be the number of atoms per unit cell, then $M/\rho$ is the volume of one mole substances. If $N$ be the Avogadro's number, then the volume of $n$ atoms is $(nM/N\rho)$, but we know the volume of the unit cell in cubic crystal is $a^3$. Then

$$a^3 = \frac{nM}{N\rho}$$

or,
$$\rho = \frac{nM}{Na^3} \tag{1.4}$$

or,
$$a = \left(\frac{nM}{N\rho}\right)^{\frac{1}{3}}.$$

For BCC crystal $n = 2$,

$$\therefore \quad a = \left(\frac{2M}{N\rho}\right)^{\frac{1}{3}}.$$

For FCC crystal $n = 4$,

$$\therefore \quad a = \left(\frac{4M}{N\rho}\right)^{\frac{1}{3}}.$$

## 1.7 MILLER INDICES

It is frequently necessary to consider a plane passing through the lattice points forming a space lattice since it may be thought of as occupying various set of parallel planes. The position and orientation of a crystal plane are determined by any three points in the plane, provided the points are not collinear. If each of the points lies on a crystal axis, the plane may be specified by giving the position of the points along the axes in terms of the lattice constants. Miller designates a plane in a crystal by three numbers (*hkl*), known as *Miller indices*. Let a particular plane of a given set have intercepts *pa*, *qb* and *rc* on the crystal axes as in Figure 1.19. Here *p*, *q* and *r* are integers and *a*, *b*, *c* are lattice constants, then

$$h : k : l = \frac{1}{p} : \frac{1}{q} : \frac{1}{r} \qquad (1.5)$$

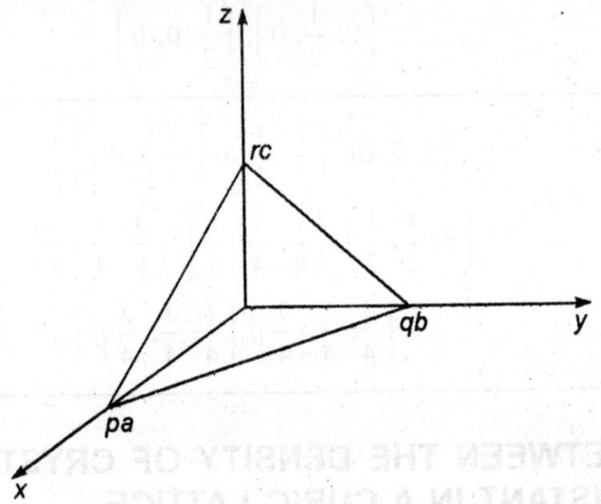


**Figure 1.19** Miller indices for planes

With the condition that *h*, *k* and *l* are the smallest integers, i.e., have no common factor greater than one, it follows that the direction cosine of the normal to the plane (*hkl*) are proportional to *h*, *k*, *l* respectively. A direction normal from the origin to the crystal plane gives the direction of the line (*hkl*). If an intercept is negative Miller indices are $\bar{h}kl$ with an overbar above the negative intercept. In a cubic latice all cube faces are equivalent in order to specify this group of planes, one can write (1, 0, 0), (0, 1, 0), (0, 0, 1), ($\bar{1}$, 0, 0), (0, $\bar{1}$, 0) and (0, 0, $\bar{1}$). Miller indices of some important plane in a cubic crystal are shown below in Fig. 1.19(a).

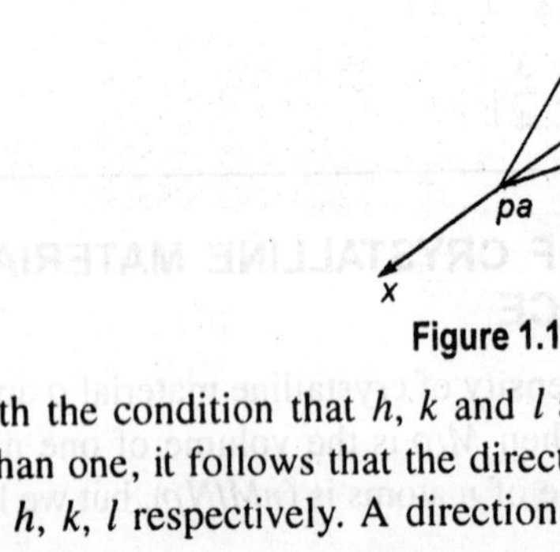


(100)

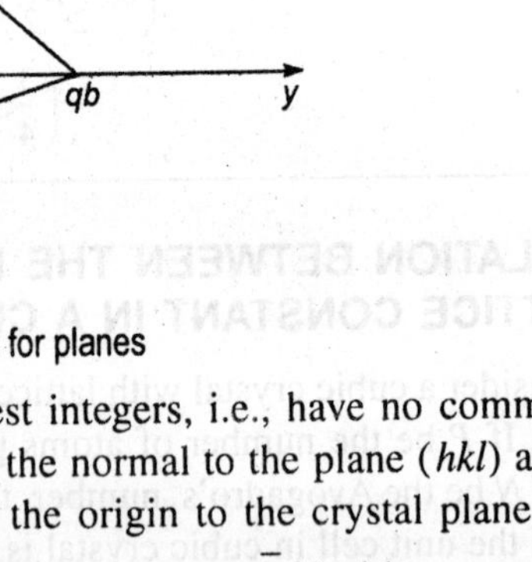


(010)

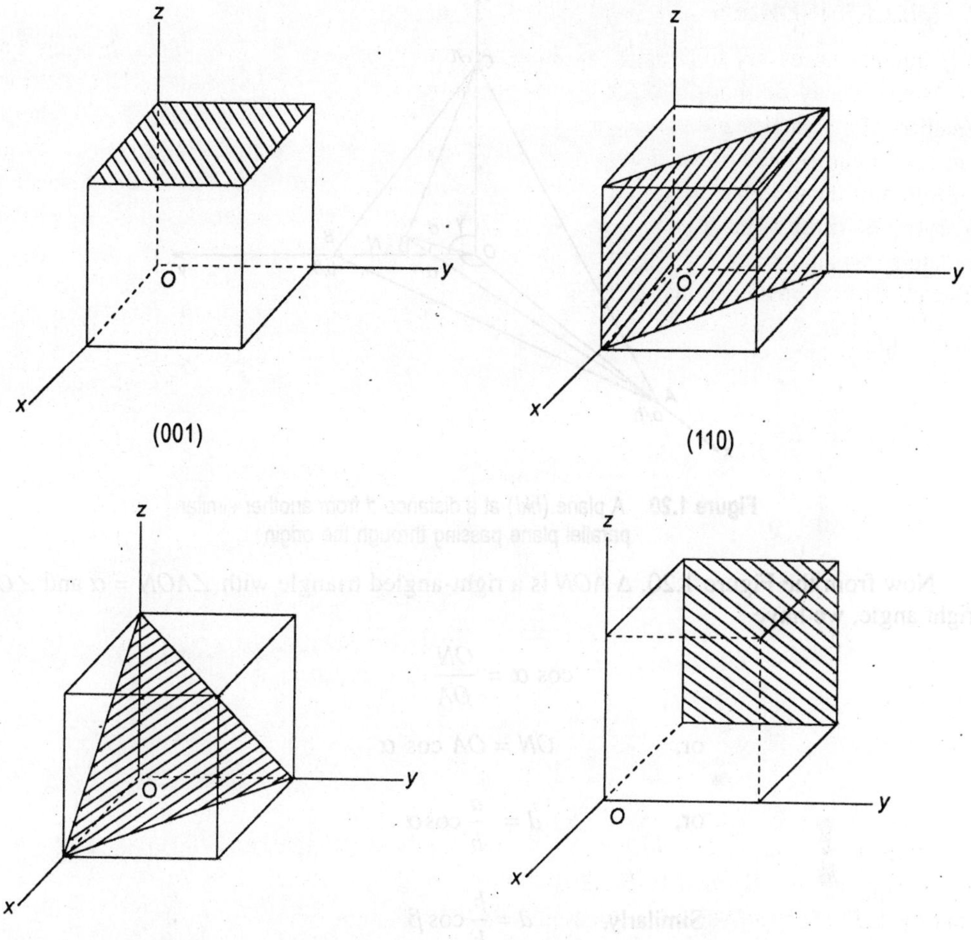


**Figure 1.19(a)** Miller indices of some plane in a cubic crystal

*Some features of a cubic system*:

(i) All the parallel planes equally spaced have the same index number (*hkl*).

(ii) A plane parallel to one coordinate axis has an intercept of infinity.

(iii) A plane passing through origin has non-zero intercept.

## 1.8 SPACING OF PLANES IN CRYSTAL LATTICE

Consider any lattice point taken as origin and the crystal axes taken in the *abc* which are orthogonal in direction. Consider a plane *ABC* (Figure 1.20) whose intercepts are $\frac{a}{h}$, $\frac{b}{k}$ and $\frac{c}{l}$ on the axes. Draw *ON* perpendicular from *O* on the plane *ABC* and let $ON = d$ and *ON* makes angles $\alpha$, $\beta$ and $\gamma$ with the axes *x*, *y*, *z* respectively, then according to the law of direction consines, we know,

$$\cos^2 \alpha + \cos^2 \beta + \cos^2 \gamma = 1. \tag{1.6}$$

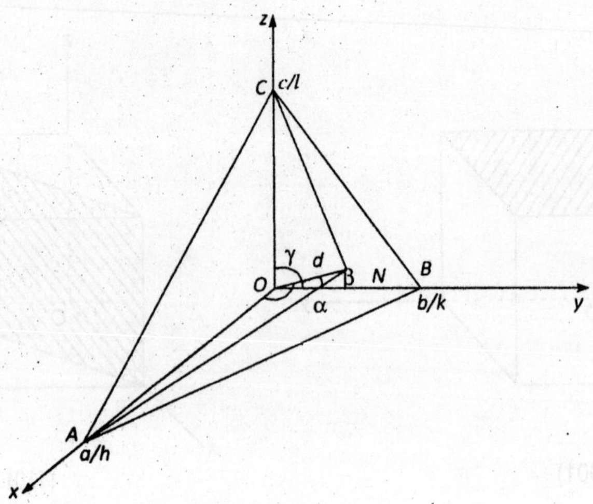


**Figure 1.20** A plane (*hkl*) at a distance *d* from another similar parallel plane passing through the origin

Now from the Figure 1.20, $\Delta\, AON$ is a right-angled triangle with $\angle AON = \alpha$ and $\angle ONA =$ 1 right angle, we have

$$\cos\alpha = \frac{ON}{OA}$$

or, $$ON = OA\cos\alpha$$

or, $$d = \frac{a}{h}\cos\alpha\,. \tag{1.7}$$

Similarly, $$d = \frac{b}{k}\cos\beta \tag{1.8}$$

and $$d = \frac{c}{l}\cos\gamma\,. \tag{1.9}$$

Now substituting the values of cos $\alpha$, cos $\beta$ and cos $\gamma$, from (1.7), (1.8) and (1.9) in equation (1.6), we have

$$d^2\left[\left(\frac{h}{a}\right)^2 + \left(\frac{k}{b}\right)^2 + \left(\frac{l}{c}\right)^2\right] = 1$$

or, $$d = \frac{1}{\left[\left(\frac{h}{a}\right)^2 + \left(\frac{k}{b}\right)^2 + \left(\frac{l}{c}\right)^2\right]^{1/2}}\,. \tag{1.10}$$

For cubic lattice, $a = b = c$, then $d$ can be written as $d_{hkl}$ as

$$d_{hkl} = \frac{a}{\left[h^2 + k^2 + l^2\right]^{\frac{1}{2}}} \tag{1.11}$$

The distance for (100) plane is

$$d_{100} = \frac{a}{[1+0+0]^{\frac{1}{2}}} = ON\,.$$

$$= a. \tag{1.12}$$

Similarly, the distance for (200) plane is

$$d_{200} = \frac{a}{\left[2^2 + 0 + 0\right]^{\frac{1}{2}}} = \frac{a}{2} \tag{1.13}$$

and the distance for (111) plane is

$$d_{111} = \frac{a}{[1+1+1]^{\frac{1}{2}}}$$

$$= \frac{a}{\sqrt{3}}\,. \tag{1.14}$$

Now consider the distance of the next plane of the same family from the origin is $OM$ and it is equal to $d_2$. The intercepts of this plane on the three crystallographic axes are

$$OA' = \frac{2a}{h},\ OB' = \frac{2a}{k} \text{ and } OC' = \frac{2a}{l}$$

and

$$\cos\,\alpha' = \frac{d_2}{OA'},\ \cos\,\beta' = \frac{d_2}{OB'} \text{ and } \cos\,\gamma' = \frac{d_2}{OC'}\,.$$

Again, $\cos^2\alpha' + \cos^2\beta' + \cos^2\gamma' = 1$

or,
$$\left(\frac{d_2}{OA'}\right)^2 + \left(\frac{d_2}{OB'}\right)^2 + \left(\frac{d_2}{OC'}\right)^2 = 1$$

or,
$$\left(\frac{d_2}{\frac{2a}{h}}\right)^2 + \left(\frac{d_2}{\frac{2b}{k}}\right)^2 + \left(\frac{d_2}{\frac{2c}{l}}\right)^2 = 1$$

or,
$$d_2^2\left[\frac{h^2}{4a^2} + \frac{k^2}{4b^2} + \frac{l^2}{4c^2}\right] = 1\,.$$

For cubic crystal, $a = b = c$.

$$\therefore \qquad d_2 = \frac{2a}{\left[h^2 + k^2 + l^2\right]^{\frac{1}{2}}} = OM\,. \tag{1.15}$$

This interplanar spacing between successive planes of Miller indices (*hkl*) is given from the equations (1.11) and (1.15) as

$$d = OM - ON = \frac{a}{[h^2 + k^2 + l^2]^{1/2}}. \qquad (1.16)$$

## 1.9 RECIPROCAL LATTICE

The Miller indices of a plane do not define a plane uniquely since all parallel planes have a common normal and hence identical Miller indices. If the length assigned to each normal is proportonal to the reciprocal of interplanar spacing in that plane, the points at the end of the normals drawn from a common origin form a lattice which is called the *reciprocal lattice*. Thus each point in a reciprocal lattice is a representative point of a particular set of planes, vector in the crystal lattice has the dimension of length and vector in reciprocal lattice has the dimension of [length]$^{-1}$. The reciprocal lattice plays an important role in the analysis of crystal structure.

**Vector algebra discussion of reciprocal lattice**

If $\vec{a}$, $\vec{b}$, $\vec{c}$ are the primitive translation vectors of a direct space lattice for a crystal forming a primitive cell, the magnitude of the volume of the cell of the crystal lattice is given by

$$V = \vec{a}\,.\,(\vec{b}\times\vec{c}\,) = \vec{b}\,.\,(\vec{c}\times\vec{a}\,) = \vec{c}\,.\,(\vec{a}\times\vec{b}\,). \qquad (1.17)$$

Choosing $\vec{A}$, $\vec{B}$, $\vec{C}$ as the primitive translation vectors of the reciprocal lattice and $\vec{A}$, $\vec{B}$, $\vec{C}$ have the property

$$\begin{array}{lll} A\,.\,\vec{a} = 2\pi & \vec{B}\,.\,\vec{a} = 0 & \vec{C}\,.\,\vec{a} = 0 \\ \vec{A}\,.\,\vec{b} = 0 & \vec{B}\,.\,\vec{b} = 2\pi & \vec{C}\,.\,\vec{b} = 0 \\ \vec{A}\,.\,\vec{c} = 0 & \vec{B}\,.\,\vec{c} = 0 & \vec{C}\,.\,\vec{c} = 2\pi \end{array} \qquad (1.18)$$

and $\vec{A}$, $\vec{B}$ and $\vec{C}$ are given by

$$\vec{A} = 2\pi\,\frac{\vec{b}\times\vec{c}}{\vec{a}.(\vec{b}\times\vec{c})},\quad \vec{B} = 2\pi\,\frac{\vec{c}\times\vec{a}}{\vec{b}.(\vec{c}\times\vec{a})},\quad \vec{C} = 2\pi\,\frac{\vec{a}\times\vec{b}}{\vec{c}.(\vec{a}\times\vec{b})}. \qquad (1.19)$$

The point $\vec{\rho}$ of a crystal lattice is given by

$$\vec{\rho} = m\vec{a} + n\vec{b} + p\vec{c}\,, \qquad (1.20)$$

where *m*, *n* and *p* are integers. Similarly, we define reciprocal lattice points or reciprocal lattice vectors $\vec{G}$ as

$$\vec{G} = h\vec{A} + k\vec{B} + l\vec{C}\,, \qquad (1.21)$$

where *h*, *k* and *l* are integers and define the coordinate of the points in the reciprocal lattice.

Now from equations (1.20) and (1.21),

$$\vec{G}.\vec{\rho} = (h\bar{A} + k\bar{B} + l\bar{C}).(m\bar{a} + n\bar{b} + p\bar{c}) \qquad (1.22)$$

Using equation (1.18), the equation (1.22) becomes

$$\vec{G}.\vec{\rho} = 2\pi\ (hm + nk + pl)$$
$$= 2\pi \cdot \text{integer.}$$
$$\therefore\ \exp(i\,\vec{G}.\,\vec{\rho}) = \exp[2\pi i\ (\text{integer})]$$
$$= 1. \tag{1.23}$$

From equation (1.21) $h$, $k$ and $l$ define the coordinates of the points of reciprocal lattice space and it means that $(h, k, l)$ in the reciprocal space corresponds to the set of parallel planes having Miller indices $(hkl)$. The reciprocal lattice concept is useful for defining Bragg's condition and the concept of Brillouin zones.

**Reciprocal lattice to the simple cubic lattice (SC)**

The primitive translation vectors of simple cubic lattice are $\vec{a}$, $\vec{b}$ and $\vec{c}$. We can write the vectors as

$$\vec{a} = a\hat{i},\ \vec{b} = a\hat{j} \text{ and } \vec{c} = a\hat{k}$$

Volume of the cubic unit cell, $V = \vec{a}\,.\,(\vec{b}\times\vec{c})$

$$V = a\hat{i}\ .\left(a\hat{j}\times a\hat{k}\right)$$
$$= a^3. \tag{1.24}$$

The reciprocal lattice vectors are:

$$\vec{A} = 2\pi\,\frac{\vec{b}\times\vec{c}}{\vec{a}\,.(\vec{b}\times\vec{c})} = 2\pi\,\frac{a\hat{j}\times a\hat{k}}{a^3} = \frac{2\pi\hat{i}}{a}. \tag{1.25}$$

Similarly,
$$\vec{B} = 2\pi\,\frac{\vec{c}\times\vec{a}}{\vec{b}\,.(\vec{c}\times\vec{a})} = 2\pi\,\frac{a\hat{k}\times a\hat{i}}{a^3} = \frac{2\pi\hat{j}}{a} \tag{1.26}$$

and
$$\vec{C} = 2\pi\,\frac{\vec{a}\times\vec{b}}{\vec{c}\,.(\vec{a}\times\vec{b})} = 2\pi\,\frac{a\hat{i}\times a\hat{j}}{a^3} = 2\pi\,\frac{\hat{k}}{a}. \tag{1.27}$$

All three reciprocal lattice vectors in simple cubic have lattice constant $2\pi/a$.

## 1.10 X-RAYS

German physicist Prof Wilhelm Conard Röntgen in 1895 discovered X-rays accidentally. He was investigating on the discharge of electricity through rarefied gases at a very high potential difference of about 40,000 volts. The discharge tube was completely enclosed within a black card-board box so that no visible or ultraviolet light could come out of it. He observed that if a piece of paper coated with barium platinocyanide was held outside the card-board it began to glow with fluorescent light and an invisible radiation along with it. This new type of radiation was named X-radiation or simply X-rays, because when this ray was discovered in 1895, the nature of the rays was unknown.

## 1.11 ORIGIN OF X-RAYS

When high speed electrons from the cathode of an X-ray tube strikes the target, the electrons penetrate the atoms of the target and remove an electron from inner shell by collision (Fig. 1.21). If an electron in $K$-shell is knocked out, then electrons from $L$, $M$, $N$ shells jump to $K$-shell as a result of which a photon energy $h\nu$ is emitted from $K$-shell and X-ray produces $K_\alpha$ lines [Fig. 1.22], when electron from $L$-shell jumps to $K$-shell. If the electrons from $M$-shell jump to $K$-shell, then $K_\beta$ lines are produced. Similarly, if electrons in $L$-shell are knocked out, then electrons from $M$, $N$, ... shells jump to $L$-shell and photon energy $h\nu$ is emitted.

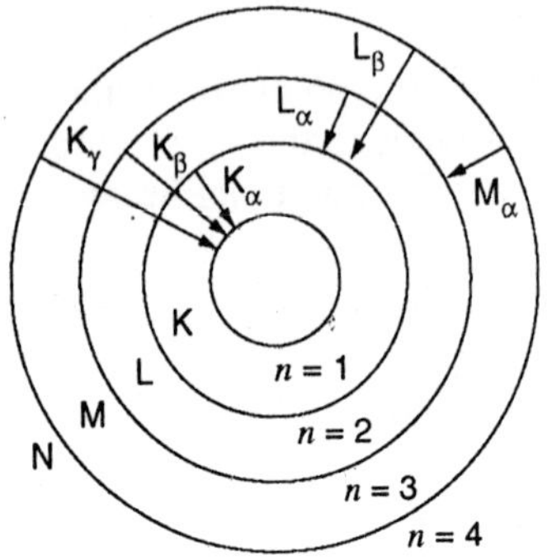


Figure 1.21 Origin of $K_\alpha$ and $K_\beta$ lines

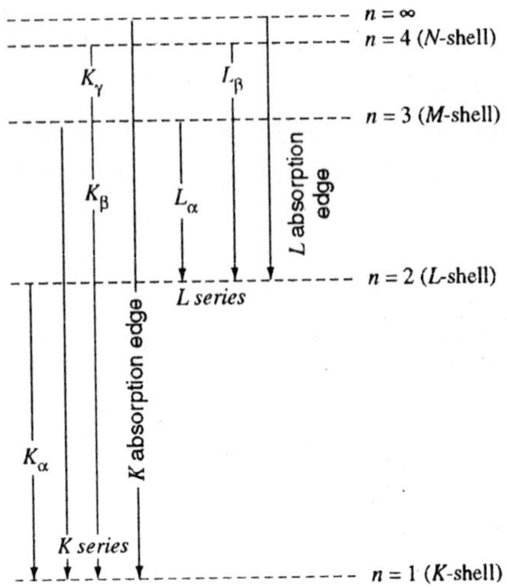


Figure 1.22 Origin of X-ray lines

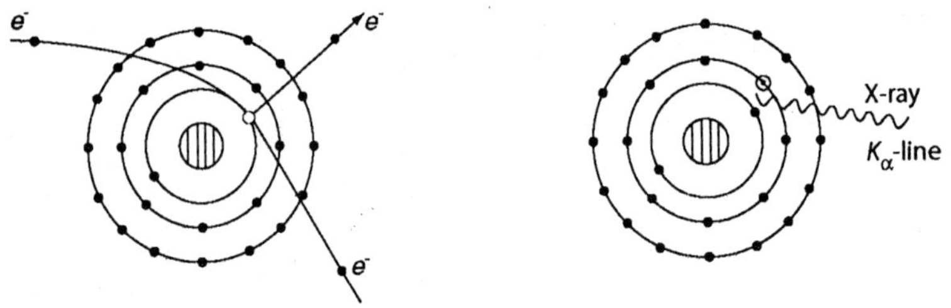


Figure 1.22(a) X-ray production

## 1.12 X-RAY SPECTRUM

### (i) Continuous X-ray spectrum

The X-ray spectrum emitted when cathode rays impinge on a target consisting of all possible wave- length, i.e., the series of bright lines superimposed on a continuous luminous background. The lines are characteristic of the substance forming the target and called continuous X-ray spectrum.

According to the quantum theory, each X-ray photon is produced as a result of particular collision of electrons. The continuous X-ray spectrum is produced because of the retardation of high velocity electrons while passing through the positively charged nucleus in the atom of the target nucleus. So the maximum energy of a photon of wavelength $\lambda$ is given by

$$E_m = h\nu_m = \frac{hc}{\lambda_{min}} = eV \tag{1.28}$$

where $h$ is the Planck's constant, $c$ is the velocity of light in vacuum, $\lambda_{mn}$ is the minimum wave- length $V$ is the potential difference between the cathode and anode.

$$\therefore \lambda_{min} = \frac{hc}{eV}. \qquad (1.29)$$

If $V$ is measured in volts, we get

$$\lambda_{min} = \frac{12413}{V(\text{volts})} \text{Å}. \qquad (1.30)$$

Thus, if $V = 40000$ volts,

$$\lambda_{min} = \frac{12413}{4000} \text{Å}$$

$$= 0.3103 \text{ Å}.$$

Shortest wavelength $\lambda_{min}$ is independent of the target nucleus.

The intensity of X-rays increases at all wavelengths as accelerating voltage is increased (Fig. 1.23).

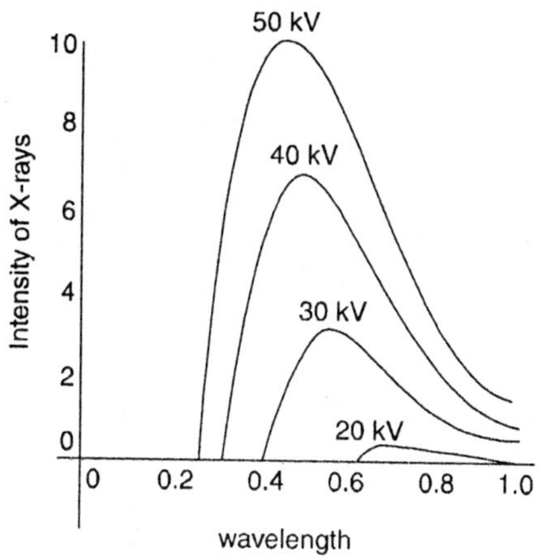


**Figure 1.23** Intensity vs. wavelength of X-rays emitted from tungsten at different voltages

### (ii) Characteristic X-ray spectrum

If a target atom is bombarded by high speed electron which is produced by high velocity potential difference, $K$-electron which is tightly bound, be knocked out. The kinetic energy of the knocked out electron is

$$\frac{1}{2}mv^2 = eV - E_K, \qquad (1.31)$$

where $eV$ represents the kinetic energy of the incident high speed electron.

The vacancy in the $K$-shell can be filled up by transition of loosely bound electrons from outer shell, i.e., $L$, $M$, $N$, etc. shells. The excess energy of the electron is taken away by a photon. The energy difference $E_K - E_L$ is the energy of the emitted X-ray photon,

$$h\nu_{KL} = E_K - E_L. \qquad (1.32)$$

So the frequency of the $K_\alpha$ line is the characteristic of the target atom.

If an electron from $M$-shell fills up the $L$-shell, $K_\beta$ lines are emitted (Fig. 1.23.) So the frequency of these lines are given by

$K_\alpha$ lines: Transition from $K \rightarrow L$ level and

$$\nu(K_\alpha) = \frac{E_K - E_L}{h}.$$

$K_\beta$ lines: Transition from $L \rightarrow M$ level

$$\nu\ (K_\beta) = \frac{E_K - E_M}{h}$$

$L_\alpha$ lines: Transition from $L \rightarrow M$ level

$$\nu\ (L_\alpha) = \frac{E_L - E_M}{h}$$

and $L_\beta$ lines: Transition from $L \rightarrow M$ level

$$\nu\ (L_\beta) = \frac{E_L - E_N}{h}.$$

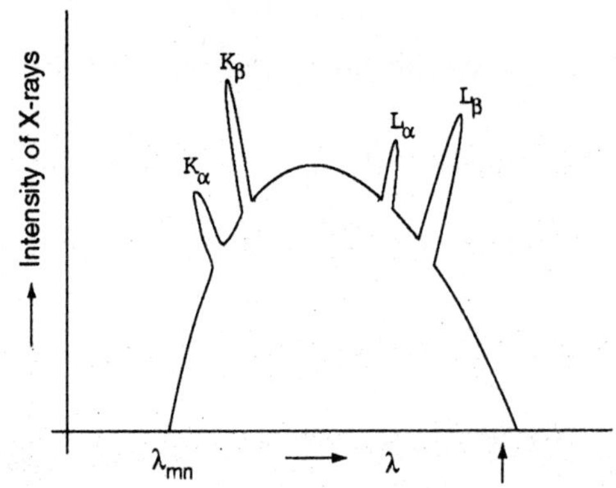


**Figure 1.24** Intensity vs. wavelength of characteristic X-ray spectrum

## 1.13 MOSLEY'S LAW

Mosley first extensively carried out the study of characteristic X-rays of various elements and found that if ν is the frequency of some particular line in the $K$ spectrum, its value in any element can be expressed by the equation

$$\nu = a\,(Z - b)^2 \tag{1:33}$$

where $Z$ is the atomic number of the element and $a$ and $b$ are constants for particular line.

$$\therefore\ \sqrt{\nu} = a_1 (Z - b) \tag{1.34}$$

where $a_1$ is a constant.

Mosley's law can be derived from Bohr's theory. Let a transition occur from state $n_1$ to $n_2$ so that energy emitted from X-ray photon is

$$h\nu = RhcZ^2 \left(\frac{1}{n_1^2} - \frac{1}{n_2^2}\right)$$

or, $$\nu = RcZ^2 \left(\frac{1}{n_1^2} - \frac{1}{n_2^2}\right)$$

$$= a^2 Z^2$$

$$\therefore\ \sqrt{\nu} = aZ, \tag{1.35}$$

where $R$ is the Rydberg's consant and $a^2 = Rc\left(\frac{1}{n_1^2} - \frac{1}{n_2^2}\right)$.

## 1.14 ABSORPTION OF X-RAYS

When a parallel X-ray beam is allowed to pass through a slab of some material part, it gets absorbed and intensity of the beam emerging from the opposite side of the slab decreases. If $I_0$ be the intensity of the beam and $I$ be the intensity after passing through a thickness $d_x$ of the slab, then $\left(-\frac{dI}{dx}\right)$ is the rate of change of intensity with thickness; negative sign indicates that $I$ decreases with increase of $x$.

$$\therefore \quad \left(-\frac{dI}{dx}\right) \propto I$$

$$\text{or,} \qquad \frac{dI}{dx} = -\mu I, \tag{1.36}$$

where μ is the characteristic constant of the material called absorption coefficient of the slab.

$$\therefore \quad \frac{dI}{I} = -\mu\, dx \tag{1.37}$$

Integrating of the equation (1.37), we get

$$\log I = -\mu x + C. \tag{1.38}$$

where $C$ is the constant of integration,

Now at $x = 0$, $I = I_0$;

$$\therefore \quad \log I_0 = C.$$

The equation (1.38) becomes

$$\log I = -\mu x + \log I_0$$

$$\text{or,} \qquad \log \frac{I}{I_o} = -\mu x.$$

$$\therefore \qquad I = I_0\, e^{-\mu x}. \tag{1.39}$$

The equation (1.39) shows that the intensity $I$ decreases with increase of $x$.

If we plot $I$ versus $x$, we get a curve as shown in Fig. 1.24.

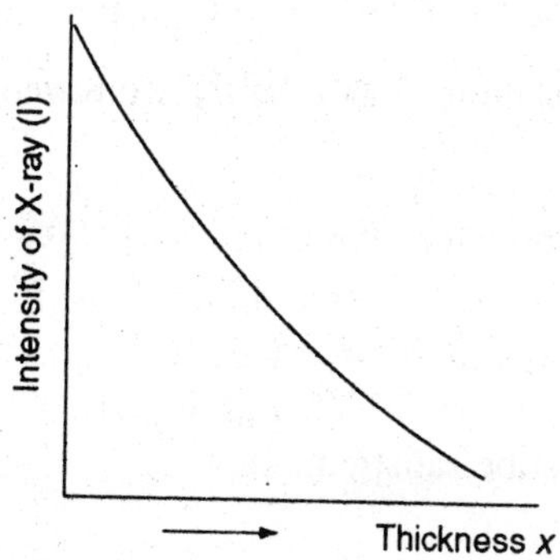


**Figure 1.25** Variation of intensity with absorber thickness

## 1.15 X-RAY DIFFRACTION: LAUE'S AND BRAGG'S EQUATIONS

In 1912 Prof Laue conceived the idea that a regular grouping of atoms in a crystal in three-dimensional array of regularly spaced atoms could serve the purpose of natural grating, In ordinary grating, spacings are all parallel and all in a plane which is not deserved in crystal actually. The crystal acts as a space grating rather than a plane grating.

Friedrich and Knipping in 1913 succeeded in diffracting X-rays by passing through a thin crystal. Diffracted rays should emerge in various directions from the crystal and would form a symmetrical pattern of spots on a photographic plate placed perpendicular to the beam. Suppose $P_1$, $P_2$, $P_3$ in Figure 1.26 are all equally spaced along a line parallel to one of the axes of the crystal. By ordinary theory of optical diffraction, wavefront $P_2Q_1$, travelling in the direction $Q_1P_2$, will give rise to diffracted wavefront $P_1Q_2$ in the direction $P_2Q_2$. We must require that the optical path difference $(P_2Q_2 - P_1Q_1)$ is a whole number of wavelengths of the radiation used as a diffracted beam, if

$$a\,(\cos\varphi - \cos\varphi_0) = p\lambda, \tag{1.40}$$

where $p$ is an integer and $P_1P_2 = P_2P_3 = a$.

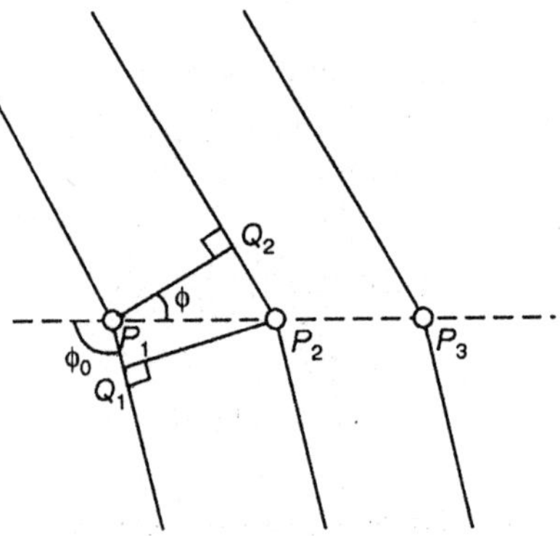


**Figure 1.26** X-rays diffraction of beams

Writing $\cos\varphi_0 = \alpha_0$ and $\cos\varphi = \alpha$, we have

$$a\,(\alpha - \alpha_0) = p\lambda. \tag{1.41}$$

If $\vec{n}_0$ and $\vec{n}$ represent unit vectors respectively in the direction of incident and scattered beams and $\vec{a}$ represents the translation from $P_1$ to $P_2$, we have

$$\vec{a}\,.(\vec{n} - \vec{n}_0) = p\lambda\,. \tag{1.42}$$

Exactly similar conditions apply for a three-dimensional crystal the atoms are equally spaced parallel to the other axes of the crystal in three dimensions. If $\vec{a}, \vec{b}, \vec{c}$ are primitive translation vectors and $\alpha_0$, $\beta_0$, $\gamma_0$ and $\alpha$, $\beta$, $\gamma$ represent respective angles between incident and scattered beams, then for diffraction the conditions which must be simultaneously fulfilled by the following equations for path difference must be satisfied.

$$\left.\begin{aligned} a\,(\cos\alpha - \cos\alpha_0) &= \vec{a}\,.(\vec{n} - \vec{n}_0) = p\lambda \\ b\,(\cos\beta - \cos\beta_0) &= \vec{b}\,.(\vec{n} - \vec{n}_0) = q\lambda \\ c\,(\cos\gamma - \cos\gamma_0) &= \vec{c}\,.(\vec{n} - \vec{n}_0) = r\lambda \end{aligned}\right\}, \tag{1.43}$$

where $p$, $q$ and $r$ are integers. The equations (1.31) are known as von Laue equations. It is to be noted that for a given value of $\lambda$ and arbitrary direction of incidence $\vec{n}_0$, it is possible to find a direction $\vec{n}$ which satisfies equations (1.31) with finite values of the integers $p$, $q$ and $r$.

The X-rays from an X-ray tube form a continuous spectrum and the crystal picks out the particular components of the radiation which are suitable for diffraction. Thus the different diffracted rays differ from wavelength, and exact interpretation of the pattern is not always easy. This point was made more clear by W. L. Bragg in 1913. Assuming $a = b = c$ for a cubic lattice we obtain from equation (1.31) by squaring and adding and using equation (1.6).

$$2a^2 (1-\cos\alpha\cos\alpha_0 - \cos\beta\cos\beta_0 - \cos\gamma\cos\gamma_0) = \lambda^2 (p^2 + q^2 + r^2) \tag{1.44}$$

If $\phi$ represents the angle of incidence between incident and scattered beams, the equation (1.32) may be written as under:

$$2(1-\cos\phi) = 4\sin^2\frac{\phi}{2} = \frac{\lambda^2}{a^2}(p^2 + q^2 + r^2) \tag{1.45}$$

The equation (1.33) is a form of the von Laue equation and is closely related to the Bragg's formula.

## 1.16 X-RAY DIFFRACTION AND BRAGG'S LAW OF CRYSTAL

Bragg considers X-ray diffraction from a crystal which is due to reflections from a set of parallel planes. Since X-rays are insignificantly absorbed in crystals, the incident beam is reflected from many mutually parallel crystal planes as in Figure 1.27. Consider parallel lattice planes of Miller indices ($hkl$) separated by a distance $d_{hkl}$ (interplanar spacing). Consider a ray $MA$ reflected at atom $A$ in the direction $AN$ and another $PB$ reflected at another atom $B$ in the direction $BQ$. From the atom $A$ draw two perpendiculars $AC$ and $AD$ on $PB$ and $BQ$ respectively. We see that the rays $MAN$ and $PBQ$ can reinforce each other in the reflected direction if their path difference ($CB+BD$) is an integer tines $\lambda$ and produces an intense spot. The condition of reinforcement is

$$(CB + BD) = n\lambda, \tag{1.46}$$

when $n = 0, 1, 2$.

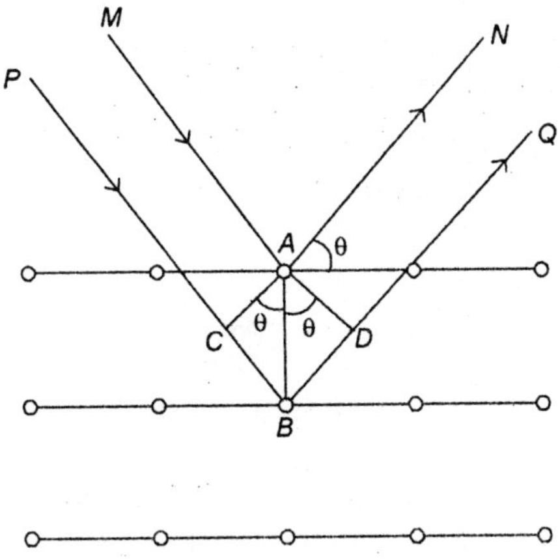


**Figure 1.27** Bragg's reflection of X-rays from the atomic planes

From Figure 1.27 $CB = BD = d_{hkl}\sin\theta$, where $\theta$ is the angle between the incident ray and the plane of reflection (glancing angle). Therefore, the equation (1.34) is

$$2d_{hkl}\sin\theta = n\lambda,\ n = 1, 2, 3, \ldots \tag{1.47}$$

The equation (1.35) is called *Bragg's equation* after W. L. Bragg who first derived it.

A beam of monochromatic X-rays incident on a crystal with arbritary angle $\theta$ is in general reflected. Since $\sin\theta \le 1$ and $d_{hkl} = 10^{-10}$ m the reflection can be observed only for $\lambda$ of the order of $10^{-10}$ m or less. For this reason X-rays are used in the experiement.

The equation (1.35) is equivalent to equation (1.31) of von Laue treatment. For simplicity we consider a simple cubic lattice and comparing with equation (1.33) we write equation (1.35) as

$$2a \sin\theta = n\lambda\,(h^2 + k^2 + l^2)^{1/2}, \tag{1.48}$$

where we assume $\varphi/2 = \theta$.

Identifying the integers $p$, $q$ and $r$ respectively as $nh$, $nk$ and $nl$ a diffracted beam defined in the von Laue treatment by the integers $p$, $q$ and $r$ may be interpreted as the $n^{th}$ order reflection from a set of planes ($hkl$) in the Bragg's theory. The order of the reflection $n$ is simply equal to the largest common fraction of the number $p$, $q$, $r$.

## 1.17 EXPERIMENTAL METHOD OF X-RAY DIFFRACTION

The Bragg's law $2d \sin\theta = n\lambda$ for X-ray diffraction requires that $\theta$ and $\lambda$. X-rays of wavelength $\lambda$ striking a three-dimensional crystal at an arbitrary angle of incidence will in general not be reflected. To satisfy the Bragg's law it is necessary *to scan either in* wavelength $\lambda$ or in angle $\theta$. The standard experimental methods are

(a) Laue's technique
(b) Rotating crystal method
(c) Powder method.

**(a) Laue's method:** In the Laue's method a single crystal is held stationary in a beam of X-rays of continuous wavelength. The crystal selects and produces diffraction corresponding to appropriate wavelength $\lambda$ for which planes exist of spacing $d$ and an incident angle $\theta$ satisfying Bragg's law. This method is convenient for determination of crystal orientation and symmetry.

Laue's X-ray camera consists of a source which produces a beam of X-rays over a wide range of wavelength (from 0.2 Å to 2 Å). A pinhole arrangement produces a well-collimated fine narrow beam. A single-crystal of specimen size not more than 1 mm in dimension is held stationary in a stand and two flat films are placed to receive transmitted diffracted and reflected beams. The film records diffraction pattern. The diffraction pattern consists of series of spots. The pattern

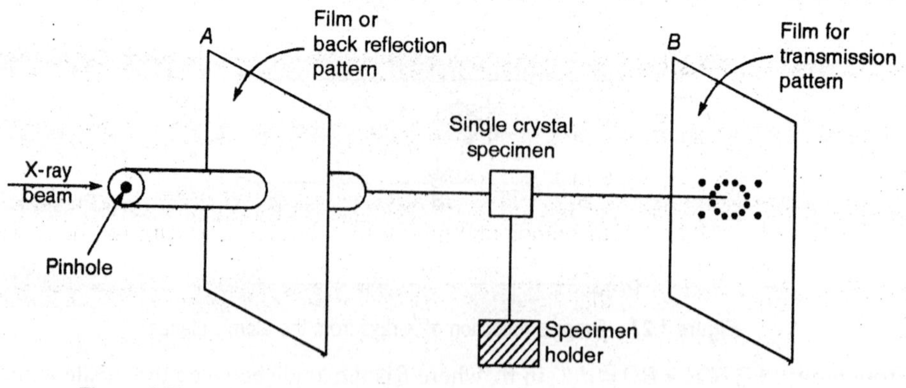


**Figure 1.28** A flat plate camera with a continuous spectrum of X-ray beam and a single crystal specimen. The camera produces Laue pattern.

will show the crystal symmetry. This method is very useful for checking the orientation of crystals in solid state physics.

**(b) Rotating crystal method:** In this method a rotating single crystal of dimension not greater than 1mm is rotated about a fixed axis in a monochromatic beam of X-rays. The variation in the angle $\theta$ brings different atomic planes into position for reflection. In this method a simple rotating X-ray camera is used. The film is mounted in a cylindrical holder concentric with a rotating spindle. The crystal is rotated slowly and during the time of exposure the successive plane passes through orientation. A diffracted beam results from a given crystal plane for which Bragg's law is satisfied for each spot produced on the film. The position on the film when developed indicates orientation of the crystal at which the spot was formed. The spacing between crystal planes is obtained from the data obtained from the spots.

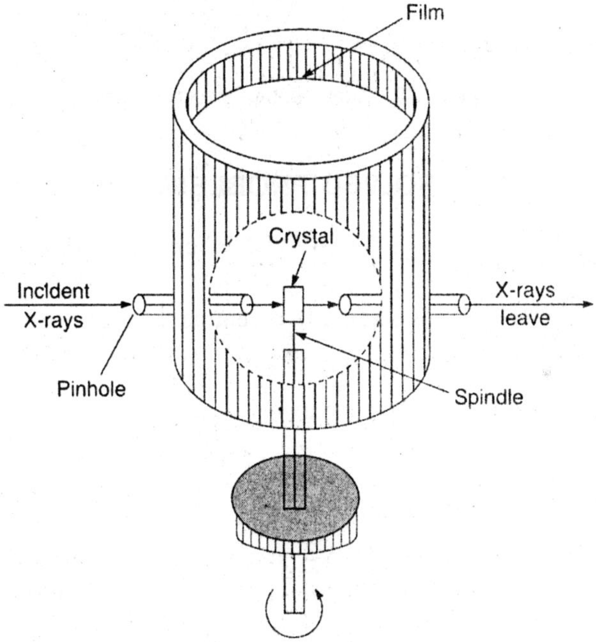


**Figure 1.29** A rotating crystal camera

**(c) Powder method:** This method of X-ray diffraction was devised by Debye and Scherrer for determination of crystal structure analysis and also used in investigation of metals and alloys. The sample used is a finely powdered specimen or fine grained polycrystalline specimen contained in a thin-walled capillary tube. The distribution of crystalline orientation will be nearly continuous. The incident parallel monochromatic X-radiation is diffracted from individual crystallites which happen to be oriented with planes making an incident angle $\theta$ with the beam satisfying Bragg's relation for wavelength $\lambda$ of X-radiation. The diffracted rays leave the specimen numerous orientations of a particular plane or a set of planes and generate a number of cones concentric with the directions with the original beam and make semivertical angle equal to $2\theta$. Different cones intercept the film in a series of concentric rings. Radii of these rings are recorded and the film can be used to find the glancing angle and hence interplanar spacing of the crystalline substance.

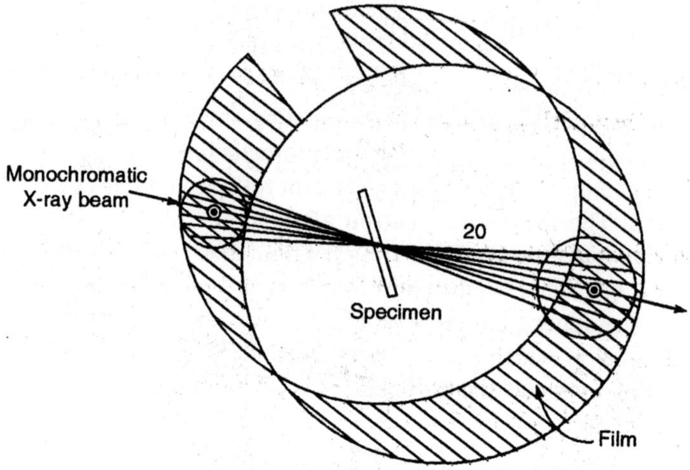


**Figure 1.30** X-ray powder diffraction camera

## SOLVED PROBLEMS

1. Show that the packing fraction in the case of FCC crystal is 0.74.

**Soln.** The total number of atoms per unit face-centred structure is $\frac{1}{8}\cdot 8 + \frac{6}{2} = 4$ [Art. 1.5(ii)]

Volume of all atoms in the unit cell $v = 4 \times \frac{4}{3}\pi r^3$,

where $r$ is the atomic radius.

Now $\because \quad (4r)^2 = 2a^2$

$\therefore$ atomic radius, $\quad r = \frac{\sqrt{2}}{4}a$.

Volume of the unit cell, $\quad V = a^3 = \left(\frac{4r}{\sqrt{2}}\right)^3$

Atomic packing fraction, $\quad \frac{v}{V} = \frac{4 \times \frac{4}{3}\pi r^3}{a^3}$

$$= \frac{4 \times \frac{4}{3}\pi r^3}{\left(\frac{4r}{\sqrt{2}}\right)^3} = 0.74 \quad \textbf{[Proved]}$$

2. Iron is BCC below 910 °C and is FCC above 910 °C. The atomic radius is the same in both structures. Calculate the ratio of densities of bulk iron in two structures.

**Soln.** Volume of the cube in BCC crystal, $V = a^3 = \left(\frac{4}{\sqrt{3}}r\right)^3$. [Art. 1.5(iii)]

Volume of all the atoms in a unit cell of BCC structure, $v = 2 \times \frac{4}{3}\pi r^3$.

Packing fraction, $$\frac{v}{V} = \frac{2 \times \frac{4}{3}\pi r^3}{a^3}$$

$$= \frac{2 \times \frac{4}{3}\pi r^3}{\left(\frac{4}{\sqrt{3}}r\right)^3}$$

$$= 0.68.$$

Packing fraction in FCC crystal $= 0.74$.

[vide Art. 1.5(ii) and also problem 1]

Hence, the ratio of packing fractions $= \frac{0.68}{0.74} = \mathbf{0.92}$.

**3.** KBr crystal has a cubic structure. If its density is $2.75 \times 10^3$ kg/m$^3$ and molecular weight is 119.01, calculate its lattice constant. **[C.U. 2000, 2006 (old syllabus)]**

**Soln.** Molecular weight of KBr, $= 119.01$ gm wt (given)

$= 119.01 \times 10^{-3}$ kg wt.

Mass of a KBr molecule, $$M = \frac{\text{Molecular weight in kg}}{\text{Avogadro number}}$$

$$= \frac{119.01 \times 10^{-3}}{6.023 \times 10^{23}} \text{ kg}$$

$$= 19.76 \times 10^{-26} \text{ kg}.$$

Number of KBr molecule per unit volume $= \frac{\text{Density}}{\text{Mass } M}$

$$= \frac{2.75 \times 10^3}{19.76 \times 10^{-26}}$$

$$= 0.139 \times 10^{29}.$$

KBr is a diatomic molecule, the number of atoms per unit volume

$$= 2 \times 0.139 \times 10^{29}$$

$$= 27.8 \times 10^{27} \text{ atoms/m}^3.$$

If $a$ be the distance between two adjacent atoms, the volume $a^3$ will contain one atom due to the cubic structure.

$\therefore$ the total number of atoms per unit volume $= \dfrac{1}{a^3}$.

Hence $$\frac{1}{a^3} = 27.8 \times 10^{27}$$

or, $$a = 3.30 \times 10^{-10} \text{ m}.$$

The lattice constant $2a = 6.6 \times 10^{-10}$ m = **6.6 Å.**

**4.** Show that for a simple cubic lattice

$$d_{100} : d_{110} : d_{111} = \sqrt{6} : \sqrt{3} : \sqrt{2}$$

where $d_{hkl}$ is the separation between adjacent $(hkl)$ parallel planes. **[C.U. 2005]**

**Soln.** The spacing $d_{hkl}$ of the plane $(hkl)$ in a simple cubic lattice of side $a$ is

$$d_{hkl} = \frac{a}{\left(h^2 + k^2 + l^2\right)^{1/2}}. \quad \text{[vide Eq 1.11]}$$

For (100) plane, $$d_{100} = \frac{a}{(1+0+0)^{1/2}} = a.$$

For (110) plane, $$d_{110} = \frac{a}{(1+1+0)^{1/2}} = \frac{a}{\sqrt{2}}.$$

For (111) plane, $$d_{111} = \frac{a}{(1+1+1)^{1/2}} = \frac{a}{\sqrt{3}}.$$

$$\therefore \quad d_{100} : d_{110} : d_{111} = a : \frac{a}{\sqrt{2}} : \frac{a}{\sqrt{3}}$$

$$= \sqrt{6} : \sqrt{3} : \sqrt{2} \quad \textbf{[Proved]}.$$

**5.** The lattice constant of a cubic crystal is 2.25 Å. Find the interplanar spacing for the set of crystallographic plane having Miller indices (100). **[C.U. 2001]**

**Soln.** The lattice constant $a = 2.25$ Å.

We know $$d_{hkl} = \frac{a}{[h^2 + k^2 + l^2]^{1/2}}\ .$$

$\therefore$ interplanar spacing $$d_1 = d_{100} = \frac{2.25}{[1+0+0]^{1/2}}\ \text{Å} = 2.25\,\text{Å}\ .$$

Let $d_2$ be the perpendicular distance from next plane parallel to (100) plane.

$$\therefore\ d_2 = \frac{2a}{[h^2 + k^2 + l^2]^{1/2}} = \frac{2 \times 2.25\ \text{Å}}{[1+0+0]^{1/2}} = 4.5\,\text{Å}\ .$$

Thus, interplanar spacing between two adjacent planes of Miller indices (100) is

$$d = d_2 - d_1 = (4.5 - 2.25)\ \text{Å}$$
$$= \mathbf{2.25\ Å}.$$

**6.** Bragg found that for a KCl crystal strong reflection from the set of planes (100), (110) and (111) is obtained at the same order for angles 5°23′, 7°37′ and 9°25′ respectively. Show that KCl crystal has a simple cubic structure.

**Soln.** From Bragg's equation we know,

$$2d \sin \theta = n\lambda.$$

For same order reflection, $n$ = constant.

$$\therefore \qquad d = \frac{n\lambda}{2 \sin \theta}$$

or, $$d \ \alpha \ \frac{1}{\sin \theta}\ .$$

Therefore, $$d_{100} : d_{110} : d_{111} = \frac{1}{\sin \theta_1} : \frac{1}{\sin \theta_2} : \frac{1}{\sin \theta_3}$$

or, $$d_{100} : d_{110} : d_{111} = \frac{1}{\sin 5°25'} : \frac{1}{\sin 7°37'} : \frac{1}{\sin 9°25'}$$

$$= 1 : 0.7076 : 0.5731. \qquad (1.37)$$

For a cubic crystal, we know,

$$d_{100} : d_{110} : d_{111} = 1 : \frac{1}{\sqrt{2}} : \frac{1}{\sqrt{3}}$$

$$= 1 : 0.7071 : 0.5773 \qquad (1.38)$$

Comparing equations (1.37) and (1.38), **we infer that the KCl crystal is cubic**.

7. The spacing between successive (100) plane in NaCl is 2.82 Å. X-rays incident on the surface of the crystal is found to give rise to first order Bragg's reflection at glancing angle 8.8°. Calculate the wavelength of X-rays. **[H.P.U. 1996]**

**Soln.** We know from Bragg's law,

$$2d \sin \theta = n\lambda.$$

Given:

$$d = 2.82 \text{ Å} = 2.82 \times 10^{-10} \text{m}$$

$$\theta = 8.8°, \quad \sin 8.8° = 0.1529$$

For first order Bragg's reflection, $n = 1$.

$$\therefore \quad \lambda = 2 \times 2.82 \times 10^{-10} \sin 8.8° \text{Å}$$

$$= 2 \times 2.82 \times 10^{-10} \times 0.1529 \text{ Å}$$

$$= 0.8623 \times 10^{-10} \text{ m}$$

$$= \mathbf{0.8623 \text{ Å}}.$$

8. A certain crystal reflects monochromatic X-rays strongly when Bragg's glancing angle (first order) is 15°. What are the glancing angles for second and third orders spectrum?

**Soln.** Bragg's law is $2d \sin \theta = n\lambda$

For first order, $n = 1$ and $\theta = \theta_1$.

$$\therefore \quad 2d \sin \theta_1 = \lambda$$

or,

$$\sin \theta_1 = \frac{\lambda}{2d}.$$

Since $\theta_1 = 15°$, $\therefore \frac{\lambda}{2d} = \sin 15° = 0.2588.$

For second order, $\theta = \theta_2$ and $n = 2$.

$$\therefore \quad \sin \theta_2 = \frac{2\lambda}{2d} = 2 \times 0.2588 = 0.5176.$$

$$\therefore \quad \theta_2 = \sin^{-1} (0.5176) = \mathbf{31.17°}.$$

For third order, $\theta = \theta_3$ and $n = 3$.

$$\therefore \quad 2d \sin \theta_3 = 3\lambda$$

or,

$$\sin \theta_3 = 3 \times \frac{\lambda}{2d} = 3 \times 0.2588 = 0.7764.$$

$$\therefore \quad \theta_3 = \sin^{-1} (0.7764) = \mathbf{50.922°}.$$

9. Sodium crystallizes in a cubic lattice with lattice constant 4.3 Å. The density of sodium is 963 kg/m³ and its atomic weight is 23. How many atoms are contained in one unit cell? What type of cubic unit cell does sodium form?

**Soln.**

Given: $a = 4.3 \text{ Å} = 4.3 \times 10^{-10} \text{ m}$

Density $\rho = 963 \text{ kg/m}^3$

Atomic weight = 23 kg/kmol.

Avogadro's number $N = 6.023 \times 10^{26}$ atoms/kmol

$$\therefore \quad a^3 = \frac{nM}{N\rho} \qquad \text{[from equation (1.4), Art. 1.6]}$$

$$\text{or,} \quad n = \frac{N\rho a^3}{M}$$

$$= \frac{6.023 \times 10^{26} \times 963 \times (4.3 \times 10^{-10})^3}{23}$$

$$= \mathbf{2.}$$

Theoretically, the effective number of atoms in a unit cell is 2 and this agrees with the body-centred cubic crystal. So sodium crystallizes in unit cell of BCC type.

**10.** Aluminium has FCC structure. Find the unit cell dimensions and atomic diameter. Given: atomic weight of Aluminium = 27 and density 2.7 gm/cm³.

**Soln.** We know from equation (1.4) in Art. 1.6,

$$\rho = \frac{nM}{a^3 N}.$$

Here $M = 27$, $N = 6.023 \times 10^{23}$, $\rho = 2.7$ gm/cm³ and $n = 4$.

$$\therefore \quad a^3 = \frac{nM}{\rho N} = \frac{4 \times 27}{2.7 \times 6.023 \times 10^{23}}$$

$$= 6.64 \times 10^{-23}$$

$$\text{or,} \quad a = 4.049 \times 10^{-8} \text{ cm}$$

$$= \mathbf{4.049\ Å}.$$

In the case of FCC structure, atomic diameter

$$2r = \frac{a}{\sqrt{2}} = \frac{4.049}{\sqrt{2}} \text{ Å} = \mathbf{2.86\ Å}.$$

**11.** Determine unit cell dimension when Bragg's angle of 45° is observed during first order reflection in a cubic crystal having Miller indices (100). Given the wavelength of the X-ray used is 2 Å.

Given: $h = 1$, $k = 0$, $l = 0$, $\lambda = 2$ Å, $\theta = 45°$ and $n = 1$.

**Soln.** Interplanar spacing, $$d_{100} = \frac{a}{(h^2 + k^2 + l^2)^{1/2}}$$

$$= \frac{a}{(1+0+0)^{1/2}} = a.$$

From Bragg's law, we have

$$2d_{100} \sin \theta = n\lambda$$

$$2a \sin 45° = 1 \times 2$$

$$\therefore \quad a = \frac{2}{2 \sin 45°} = \frac{2\sqrt{2}}{2} = \mathbf{1.41\ Å}.$$

**12.** Calculate interplanar spacing $d$ of the plane (111), which is a simple cubic lattice of side $a$. **[WBUT 2003]**

**Soln.** Interplanar spacing $d$ is given by

$$d_{111} = \frac{a}{(h^2 + k^2 + l^2)^{1/2}}$$

$$= \frac{a}{(1+1+1)^{1/2}}$$

$$= \frac{a}{\sqrt{3}}.$$

**13.** The spacing of the planes in a crystal is 1.2 Å and the angle for the 1st order Bragg's reflection is 30°. Determine the energy of the X-rays in eV. **[WBUT 2005]**

**Soln.** From Bragg's law, we have

$$2d \sin \theta = n\lambda.$$

Given: $d = 1.2 \times 10^{-8}$ cm, $\theta = 30°$ for $n = 1$.

$$\therefore \quad \lambda = \frac{2d \sin \theta}{n}$$

$$= \frac{2 \times 1.2 \times 10^{-8} \times \sin 30°}{1}$$

$$= 1.2 \times 10^{-8} \text{cm}.$$

We know, the energy of the X-ray is given by

$$E = h\nu = \frac{hc}{\lambda}.$$

Here $h = 6.62 \times 10^{-27}$ erg. see, $c = 3 \times 10^{10}$ cm

$$\therefore \quad E = \frac{6.62 \times 10^{-27} \times 3 \times 10^{10}}{1.2 \times 10^{-8}} \text{ ergs}$$

$$= \frac{6.62 \times 10^{-27} \times 3 \times 10^{10}}{1.2 \times 10^{-8} \times 1.6 \times 10^{-12}} \text{ eV}$$

$$= \mathbf{10.344\ keV}.$$

**14.** A beam of X-rays of wavelength 0.842 Å is incident on a crystal at a glancing angle of 8°35′ when first order Bragg's reflection occurs. Calculate the glancing angle for the third order reflection.

**[WBUT 2002, 2007]**

**Soln.** We know from the Bragg's law,

$$2d \sin \theta = n\lambda.$$

From the problem we have for first order reflection,

$$2d \sin 8°35' = 1 \times 0.842 \times 10^{-8} \quad (1.39)$$

and for third order reflection,

$$2d \sin \theta_3 = 3 \times 0.842 \times 10^{-8}. \quad (1.40)$$

From equations (1.39) and (1.40), we have

$$\frac{2d \sin \theta_3}{2d \sin 8°35'} = \frac{3 \times 0.842 \times 10^{-8}}{1 \times 0.842 \times 10^{-8}}$$

or, $\sin \theta_3 = 3 \times \sin 8°35'$.

$\therefore$ $\boldsymbol{\theta_3 = 26°36'}$.

**15.** The distance between (100) planes in a body-centre of a cubic structure is 0.232 nm. What is the size of the unit cell ?

**[WBUT 2006]**

**Soln.** If $r$ be radius of the atom in 3 cc crystal,

then $r = \frac{\sqrt{3}}{4} a$, where $a$ is the lattice constant.

For BCC crystal, $d_{100} = \frac{a}{2}$.

$\therefore a = 2d_{100} = 2 \times 0.232$ nm

$= \mathbf{0.464}$ **nm**.

$\therefore r = \frac{\sqrt{3}}{4} \times 0.464$ nm

$= \mathbf{0.2009}$ **nm**.

**16.** If the energy levels of KLM state of Pt are 78 keV, 12 keV and 3 keV respectively, then calculate the wavelength of $K_\alpha$ and $K_\beta$ lines emitted from Pt. [$h = 6.626 \times 10^{-34}$ Js, $c = 3 \times 10^8$ m/s]

**[WBUT 2008]**

**Soln.** Energy of $K$ level $E_K = 78$ keV $= 78 \times 10^3 \times 1.6 \times 10^{-19}$ J

Energy of $L$ level $E_L = 12$ keV $= 12 \times 10^3 \times 1.6 \times 10^{-19}$ J

Energy of $M$ level $E_M = 3$ keV $= 3 \times 10^3 \times 1.6 \times 10^{-19}$ J

The wavelength of $K_\alpha$ lines,

$$\lambda_\alpha = \frac{hc}{E_K - E_L} = \frac{6.626 \times 10^{-34} \times 3 \times 10^8}{78 \times 10^3 \times 1.6 \times 10^{-19} - 12 \times 10^3 \times 1.6 \times 10^{-19}} = 0.1882 \times 10^{-10} \text{ m} = \mathbf{0.1882\ Å}.$$

The wavelength of $K_\beta$ -lines $\lambda_\beta = \frac{hc}{E_K - E_M} = \frac{6.626\times10^{-34}\times3\times10^8}{78\times10^3\times1.6\times10^{-19}-3\times10^3\times1.6\times10^{-19}}$

$$= 0.1656 \times 10^{-10}\ \text{m} = \mathbf{0.1656\ \text{Å}}.$$

**17.** The energy gap in Ge is 0.72 eV. Find the maximum wavelength of radiatin which can cause transition from the valence band to conduction band. **[WBUT 2007]**

**Soln.** The band gap energy $E_g$ is given by

$$E_g = \frac{hc}{\lambda_{max}}$$

or, $\lambda_{max} = \frac{hc}{E_g} = \frac{6.62\times10^{-34}\times3\times10^8}{0.72\times1.6\times10^{-19}}\text{m}$

$$= \mathbf{17.24\times10^{-7}\ m}.$$

## REVIEW QUESTIONS

1. (a) What is the difference between a crystal and an amorphous solid? **[C.U. 1996, 2005]**
   (b) Cite one example for each. **[C.U. 1996]**
2. Define crystal lattice, unit cell, primitive cell and crystal translation vector.
3. All crystals can be grouped under seven crystal systems of which the cubic system is one. How many crystal classes are there and how many of them belong to cubic or isometric system? **[C.U. 2006]**
4. Write down Laue's equation for X-ray diffraction from a crystalline solid and obtain Bragg's law from Laue's equations. How does the Laue's approach differ from the Bragg's approach? **[C.U. 1996, 2001, 2004]**
5. State what do you understand by the terms (i) crystal lattice (ii) unit cell (iii) coordination number (iv) packing factor (v) atomic radius. Calculate the atomic radii in the case of FCC and BCC lattices.
6. What are Miller indices? Show that in a cubic crystal the spacing between consecutive parallel planes of Miller indices (*hkl*) is given by

$$d_{hkl} = \frac{a}{(h^2+k^2+l^2)^{1/2}}.$$

**[P.U. 1995, A.U. 1993, C.U. 1994, 1999]**

7. Describe the crystal structures of the following materials:
   (a) NaCl (b) Diamond.
8. Describe seven crystal systems with necessary diagrams.
9. What is unit cell? **[WBUT 2002]**
10. What are Miller indices? Find the packing factor for simple cubic structure. **[WBUT 2003]**

11. Can X-ray be diffracted by a plane grating? **[WBUT 2004]**

12. (a) Write down short note on Bravais lattice and Miller indices. **[WBUT 2002]**

    (b) Explain briefly the determination of crystal structure using the powder diffraction method. Give a schematic diagram of the experimental set-up. **[C.U. 2002]**

13. What do you mean by crystalline and amorphous solids? Cite one example for each. **[C.U. 2003]**

14. How many Bravais lattices are possible in three dimensions? Of these how many belong to the cubic lattice system? Name them. **[C.U. 2006]**

15. What are the Miller indices? Derive an expression for the interplanar separation of the $(h, k, l)$ planes of a simple cubic lattice.

16. Discuss briefly the difference in the nature of bonding found in diamond and KBr crystals. **[C.U. 2000]**

17. Explain and derive Bragg's law of X-ray diffraction from a crystal. **[WBUT 2005]**

18. Find out the packing fraction for FCC structure. Find out the expression for the separation between the planes with Miller indices *hkl* and lattices constant *a* for a simple cubic structure. **[WBUT 2008]**

19. Can X-ray be diffracted from a single slit of width 0.1 mm? Explain. **[WBUT 2005]**

## PROBLEMS

1. Show that the packing fraction in the case of simple cubic lattice is 0.52.

2. NaCl crystal has a cubic structure. If its density is 2.163 gm/cm$^3$ and its molecular weight is 58.45, calculate its lattice constant. **[C.U. 2004, Burd. U. 1996]**
   [**Ans.** 2.82 Å]

3. The lattice constant of an FCC lattice is 6.38 Å. Find (i) the distance between a corner atom at the base and the atom at the centre of the top face and (ii) the largest distance between two atoms in the cubic cell. [**Ans.** (i) 7.81 Å (ii) 11.05 Å]

4. Show that for a simple cubic lattice,

$$d_{100} : d_{110} : d_{111} = \sqrt{6} : \sqrt{3} : \sqrt{2}.$$

5. Calculate the Bragg's angle for X-rays having wavelength 1.54 Å in different order 1, 2 if the interplanar distance is 2.67 Å. **[A.U. 1994]**
   [**Ans.** $\theta_1 = 16.76°$, $\theta_2 = 35.22°$]

6. Copper has FCC structure and its atomic radius is 1.278 Å. Calculate its lattice constant. Take the atomic weight of copper as 63.5. [**Ans.** 7.22 Å]

7. For a cubic lattice, calculate the distance of (123) and (234) planes from a plane passing through the origin. **[H.P.U. 1993]**

$$\left[\textbf{Ans. } \text{(i) } d_{123} = \frac{a}{\sqrt{14}},\ d_{234} = \frac{a}{\sqrt{29}}\right]$$

8. KCl is a cubic crystal having density of 1.98 gm/cm$^3$.

(a) Find the distance between adjacent atoms in this crystal. **[Ans.** 3.14 Å]
(b) Find the distance from one atom to the next of the same kind. **[Ans.** 6.28 Å]

**9.** Copper has FCC structure and atomic radius 0.1278 nm. Calculate its density and interplanar spacing for (321) planes. Take the atomic weight of copper as 63.5. **[WBUT 2005]**

**[Ans.** 8.98 gm/cc, 0.966 Å]

**10.** Show that in a cubic crystal of side $a$ the spacing between consecutive planes of Miller indices (*hkl*) is

$$d_{hkl} = \frac{a}{[h^2 + k^2 + l^2]^{1/2}}$$

**11.** Determine unit cell dimension when Bragg's angle of 45° is observed during first order reflection in a cubic crystal having Miller indices (111). Given the wavelength of the X-ray used is 2 Å. **[Ans.** 2.45 Å]

# 2
# *Bonding in Solids*

## 2.0 INTRODUCTION

A solid represents a particular form of condensed matter and it has a geometrical shape. Solids are classified according to their structure as *crystalline*, *paracrystalline* and *amorphous.* Crystalline materials can be classified in many ways: (i) according to the lattice structure (ii) according to their electrical properties, mechanical properties, thermal properties, etc. (iii) according to interatomic binding forces.

The attractive electrostatic interaction between negative charges of electrons and the positive charges of nuclei are the cause of cohesion of solids. Magnetic force has a weak effect on cohesion of solids, and gravitational force is negligible. A crystal is stable when the total energy of the atoms or molecules is less than the total energy of individual atoms or molecules. The difference between free atom energy and crystal energy is known as cohesive energy. The large difference in the cohesive energy indicates that chemical bonds of very different nature are involved. According to the chemical binding forces the crystals are classified into five principal types. These type are:

(i) Molecular (A, Kr, Xe, etc.)

(ii) Ionic (NaCl, CaCl, etc.)

(iii) Metallic (Na, Ni, etc.)

(iv) Covalent (Ge, Si, etc.)

(v) Hydrogen bonded crystals.

Line of demarcation is not rigidly drawn because some materials may belong to more than one class such as ZnS which can be considered as a mixture of ionic and covalent bindings.

## 2.1 MOLECULAR CRYSTALS

Solids consisting of lattice of molecules such as hydrogen, methane, or more complex organic compound bound by weak van der Waals force and, therefore, retaining much of their individualities. Cohensive energy is defined as the energy released when the atoms or molecules at very low temperature are allowed to condense into solid phase. Inert gas solids are the example of molecular solids. Interatomic force is responsible for binding the atoms or molecules and it is also known as Johannes van der Waals binding or known as the London interaction. Johannes van der Waals attraction is the force between two atoms having closed electronic shell structure. This is due to the mutual polarization of atoms. Even if an atom or molecule has no average dipole moments it has in general an instantaneous fluctuating dipole moment from the instantaneous positions of the electrons in the orbit. An instantaneous dipole moment of magnitude $p_1$ on one atom produces an electric field $E = \frac{2p_1}{R^3}$, where $R$ is the distance of separation of the two atoms. This field well induces an instantaneous dipole moment $p_2 = \frac{2\alpha p_1}{R^3}$ on the second atom, $\alpha$ is known as electronic polarizability. The mutual potential energy of the two atoms of the form:

$$\phi(R) = -\frac{2p_1 p_2}{R^3} = -\frac{4\alpha p_1^{\,2}}{R^6} = -\frac{C}{R^6}. \tag{2.1}$$

Negative sign indicates that the interaction is attractive in nature. Because of the $R^{-6}$ dependence the mutual potential energy increases rapidly with shorter distances. The typical value of $C$ in equation (2.1) is given below. In SI system equation (2.1) becomes

$$\phi(R) = -\frac{\alpha p_1^2}{8\pi^2\varepsilon_0^2}\frac{1}{R^6} = -\frac{C}{R^6}, \tag{2.2}$$

where $C = \frac{\alpha p_1^2}{8\pi^2\varepsilon_0^2}$.

**Table 2.1**

| *Element* | Kr | Li | Cu | Na | K |
|---|---|---|---|---|---|
| $C \times 10^{41}$ Jm$^{-6}$ | 0.01 | 0.07 | 0.41 | 1.68 | 24.3 |

$\phi(R)$ is more attractive for shorter distance, i.e., $R$ decreases and it tends to make the solid collapse. This tendency is balanced by repulsive (Coulomb) potential which arises due to electronic clouds of the atoms. The total interatomic potential energy $\phi(R)$ may be put herebelow in the form known as *Leonard- Jones potential*:

$$\phi(R) = \phi_0\left[\left(\frac{\sigma}{R}\right)^{12} - \left(\frac{\sigma}{R}\right)^{6}\right], \tag{2.3}$$

where $\phi_0$ and $\sigma$ are the parameters, the value of which can be obtained from measurements made in the gas phase of the crystal.

The van der Waals bonding is weakest bond and the structures formed by this bonding are poorly stable, quite volatile and have low melting points. These types of crystals are insulators and transparent for electromagnetic radiation. The most of the molecular crystals are in general diamagnetic.

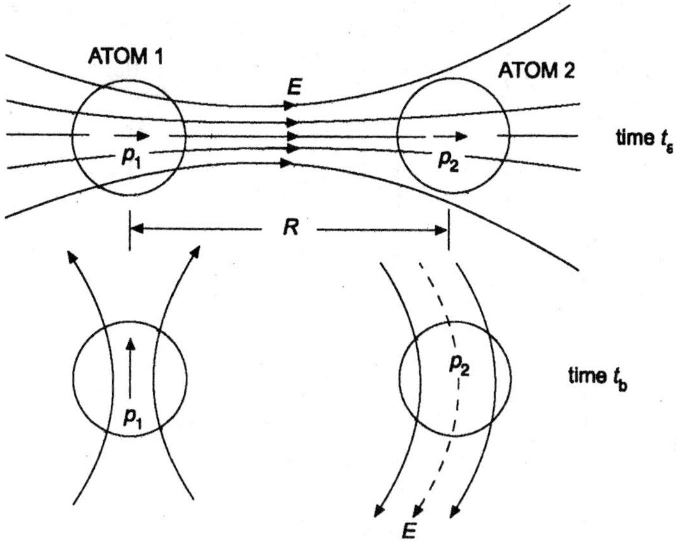


**Figure 2.1** Two atoms separated by a distance $R$ are shown. Dipole moment of magnitude $p_1$ on atom 1 will produce field $\vec{E}$

## 2.2 IONIC CRYSTALS

Ionic crystals are made of strongly positive and strongly negative ions. Their ions arrange themselves with Coulomb attraction. NaCl, KBr, LiF, MgO, etc., are examples of ionic crystals. In solid state, ionic crystals can be described as an ensembles of hard sphere which try to occupy a minimum volume while minimising electrostatic energy at the same time. There are no free electrons, for example, NaCl may be considered to be built up of $Na^+$ and $Cl^-$ ions. The electrostatic interaction between positive ion and negative ion is the source of cohesive energy. The *cohesive energy* of the NaCl crystals is to a large extent determined by the Coulomb interaction between heteropolar ions. The electronic configuration of the ions is basically inert gas structure. For example, in NaCl $Na^+$ has an electronic configuration $1s^2 2s^2\ 2p^6$ which is the same as inert gas Neon whereas $Cl^-$ has $1s^2 2s^2\ 2p^6 3s^2 3p^6$ which is the same as Argon. The charge distribution on each ion on ionic crystal will be approximately spherical symmetry. Interatomic binding forces arise from two opposing contributions.

1. *Coulomb interatomic potential energy* of the form $\pm e^2/R$ between two atoms at a separation $R$, (+) sign for like ions $Na^+$–$Na^+$ or $Cl^-$–$Cl^-$ and (–) sign for unlike ions $Na^+$–$Cl^-$.
2. There is interaction due to the overlap of the electronic shell of neighbouring atoms and this is of the form $Ae^{-R/b}$, $A$ and $b$ being empirical constants.

The contribution of the total energy of all ions due to their mutual Coulomb potential is

$$V_M(R) = N \sum \pm \frac{e^2}{R_{ij}} = -\frac{N\alpha e^2}{R}, \qquad (2.4)$$

where $R_{ij}$ is the separation between the $i$-th and the $j$-th atoms in ionic crystal and $R$ is the nearest neighbour interatomic distance. $V_M(R)$ is called the *Madelung energy* and $\alpha$ the *Madelung constant* and it is a dimensionless quantity.

Ionic crystals are strong and rigid. They are crystalline in nature. They are normally transparent for visible light and exhibit a strong absorption peak in infrared region. Ionic crystals are diamagnetic. Ionic crystals are soluble in ionizing solvent like water. They have very high melting point and possess high latent heat of fusion. Ionic crystals are insulator and their electrical conductivity is very low at ordinary temperature but increases with the rise of temperature.

A few example of ionic crystals are sodium chloride, caesium chloride, magnesium chloride, potassium chloride, lithium fluoride, zinc sulphide, etc.

## 2.3 METALLIC CRYSTALS

Large numbers of free electrons are present in metallic crystals and have high degree of mobility. The cohesive energy of metals is provided essentially by the Coulomb interaction of positive ions and the negative charge of the conduction electrons. Most metallic structures tend to be closed-packed due to the smallness of the metallic ion cores. The most common metallic structures are FCC, BCC and HCP. Metals have relatively low melting point and boiling point but transition metals fall in the upper limit, i.e., they have extremely high melting and boiling points. Metallic crystals have high electrical and thermal conductivities due to the large number of free electrons. Metals are opaque to light since the light energy is absorbed by free electrons. The spin and magnetic moment of the conduction electrons in metals are evenly paired in the absence of magnetic field except in ferromagnetic metals. This balance of magnetic moment is disturbed by applied field.

## 2.4 COVALENT CRYSTALS

In covalent crystals neighbouring atoms share their valence electrons under the formation of strong homopolar or *covalent bonds.* The covalent bond has strong directional properties. The binding of molecular hydrogen is a simple example of covalent bond. A covalent bond is a result of minimising the energy by sharing two electrons with antiparallel spins between two neighbouring nuclei. The classic example of covalently bonded crystals are the Group IV semiconductors, viz., C, Si, Ge. The covalent bonded crystals involve sharing between atoms of an even number of electrons, two per single bond. Covalent crystals are very hard and are difficult to cleave but brittle. These crystals have high refractive index and high dielectric constant. These crystals are transparent for long wavelength radiation but opaque to shorter wavelength. In covalent crystals electrical conductivity increases with the rise of temperature. Pure covalent crystals are diamagnetic but addition of impurity to these crystals shows evidence of paramagnetism.

## 2.5 HYDROGEN BONDED CRYSTALS

Neutral hydrogen atom has one electron and it should form a covalent bond with only one other atom. Hydrogen bonding is important in a few crystals notably in ice with its lone electrons. A hydrogen atom usually forms a single covalent bond with an electronegative atom. In the hydrogen bond the hydrogen atom loses the electrons to another atom in the molecule the bare proton forms the hydrogen bond. Hydrogen bonding occurs with most electronegative ions, viz., oxygen, nitrogen, fluorine. In water hydrogen links pair of oxygen atoms. Under certain condition hydrogen atom is attracted by strong forces to two atoms and hydrogen bond is formed. Ice (solid $H_2O$)

is an example of hydrogen bonded crystals. Hydrogen bonded molecule is polarised and acts like a permanent electric dipole moment and such dipoles are attracted to one another by Coulomb forces as

$$V \propto \frac{1}{r^2}. \tag{2.5}$$

*Hydrogen bonded crystals* are strong and have higher melting and boiling points than those of molecular crystals.

## 2.6 DIFFERENCES BETWEEN IONIC, COVALENT, AND METALLIC BONDING IN SOLIDS

| | Ionic Bond | Covalent Bond | Metallic Bond |
|---|---|---|---|
| Bonding forces | Ionic bonds are made of positive and negative ions | Covalent bond formed by the sharing of valence electrons | Metallic bond formed by cohensive energy of metals is provided by the Coulomb interaction of positive ions and conduction electrons |
| Melting and boiling points | Intermediate | High melting and boiling points | Low melting and boiling points |
| Colour | Wide variety of colours | White to grey | Usually grey |
| Conductivity | Very low conductivity | Low conductivity | High electrical conductivity |
| Lustre | Transparent or coloured | Transparent for long wavelength but opaque to shorter wavelength | Opaque to light since light energy is absorbed by free electrons |
| Examples | NaCl, KCl crystal | Germanium, silicon | Al, K, Au, Ag |

## REVIEW QUESTIONS

1. Describe how crystals have been classified according to their bonding. **[A.U. 1994]**
2. Write down the nature of bonding in diamond from carbon and $CO_2$ and explain briefly the mechanism of such bonding. **[C.U. 1998]**
3. Describe the difference between ionic, covalent and metallic bonding in solids. **[C.U. 1983]**
4. Explain the origin of the van der Waals interaction between two atoms of the inert gas in the solid state. **[Burd. U. 1996]**
5. What are ionic crystals? Explain the formation of ionic crystals. What is Madelung costant?
6. Write a note on:

   (a) Molecular crystals (b) Ionic crystals

   (c) Metallic crystals (d) Covalent crystals

# 3

# *Free Electron Theory of Solids*

## 3.0 INTRODUCTION

Following are the characteristics of metal:

1. Metal has a high electrical conductivity above certain temperature.
2. Metal has high thermal conductivity at high temperature.
3. The ratio of thermal conductivity ($K$) and electrical conductivity ($\sigma$) at a particular absolute temperature is the same for all metals.
4. Metal has high optical absorption.
5. Metal has temperature independent carriers concentration.

The metallic properties are determined by metallic bond (vide Art. 2.3).

## 3.1 FREE ELECTRON MODEL

Some important physical properties are explained by *free electron theory* of metals. The classical *free electron theory* was first proposed by Drüde in 1900 to explain both electrical and thermal conductivity of metals. Drüde assumed in the free electron theory that the valence electrons in the metal were able to move freely in the metal-like particles of ideal gas molecules and obey the assumptions of the kinetic theory. Since the electrons collide and are in equilibrium like the atoms of the metal, they will have, from the law of equipartition of energy, the same energy per degree of freedom as the atom inside the metal. The valence electrons of the atoms are the conductors of electricty in the metal which are called *conduction electrons*. In this mode all electron-electron and electron-ion interactions are neglected.

Lorentz extended the *free electron model* by assuming Maxwell-Boltzmann distribution of electron velocity and hence known as *Drüde-Lorentz theory*. Ohm's law and Wiedemann-Franz law were derived from Drüde-Lorentz theory which states that the ratio of the thermal and electrical conductivities for all metals is directly proportional to absolute temperature.

## 3.2 DEDUCTION OF WIEDEMANN-FRANZ LAW

In deducing Wiedemann-Franz Law following assumptions are made: (1) The free electrons are responsible for both electrical and thermal conductions in metals. (2) The free electrons are obeying Maxwell's distribution law of velocities and equipartition law.

Let us consider the conduction of electricity in a metallic wire. Let $E$ be the electric intensity along the wire. The force $F$ exerted on the elctron is $F = ma = eE$, where $m$ is the mass of the electron, $a$ the acceleration of the electron, $e$ is the charge of the electron.

$\therefore$ the acceleration,
$$a = \frac{eE}{m}: \tag{3.1}$$

The initial velocity of the electron is zero and the velocity $v$ after time $t$ is

$$v = 0 + at = \frac{eE}{m}t. \tag{3.2}$$

If $\lambda$ be the mean free path of the electron and $\bar{c}$ the mean velocity of the electron at the absolute temperature $T$, then the time, $t = \frac{\lambda}{\bar{c}}$

$\therefore$ the average drifts velocity,
$$v = \frac{1}{2}(0 + at)$$
$$= \frac{1}{2}\frac{eE}{m}\frac{\lambda}{c}. \tag{3.3}$$

The current per unit area,
$$i = ne\,\vec{v}$$
$$= ne.\frac{1}{2}.\frac{eE}{m}\frac{\lambda}{\bar{c}}$$
$$= \frac{1}{2}\frac{ne^2\lambda}{m\bar{c}}E. \tag{3.4}$$

The *electrical conductivity* $\sigma$ of unit length and unit area is given by equation (3.4) as

$$\sigma = \frac{i}{E} = \frac{1}{2}\frac{ne^2\lambda}{m\bar{c}}. \tag{3.5}$$

From equipartition law,
$$E = \frac{1}{2}m\bar{c}^2 = \frac{3}{2}kT, \tag{3.6}$$

where $k$ is the Boltzmann's constant.

$$\therefore \quad \sigma = \frac{1}{2}\frac{ne^2\lambda\bar{c}}{3kT} = \frac{ne^2\lambda\bar{c}}{6kT}. \tag{3.7}$$

But thermal conductivity $K$ for electron gas is given by

$$K = \frac{1}{3}n\bar{c}\lambda\frac{dE}{dT}. \tag{3.8}$$

From equation (3.6), we have

$$\frac{dE}{dT} = \frac{3}{2}k. \tag{3.9}$$

From equations (3.8) and (3.9), the value of $K$ can be written as

$$K = \frac{1}{3} n\bar{c}\lambda . \frac{3}{2} k . \tag{3.10}$$

From equations (3.5) and (3.10), we get

$$\frac{K}{\sigma} = 3\left(\frac{k}{e}\right)^2 T \tag{3.11}$$

or,
$$\frac{K}{\sigma T} = \text{constant},$$

which gives the laws of both Wiedemann-Franz and Lorentz.

## 3.3 SOMMERFELD'S MODEL OR QUANTUM THEORY OF FREE ELECTRONS

Classical free electron theory of Drüde and Lorentz failed to explain (i) the discrepancy between observed and predicted specific heats of metals and (ii) paramagnetism of metals which is independent of temperature.

An electron in a metal or in any solid finds itself in the field of all nuclei and all other electrons. The potential energy for such an electron may, therefore, be expected periodic. Sommerfeld assumed in his model that free electrons which give rise conductivity find themselves in a potential which is constant everywhere inside the metal. Since there is no electron emission at room temperature it seems evident that potential energy at rest inside the metal must be lower than that of an electron at rest outside the metal. One thus arrives at a physical model in which interior of the metal is represented by a potential energy box as in Figure 3.1 bounded by infinite energy barrier.

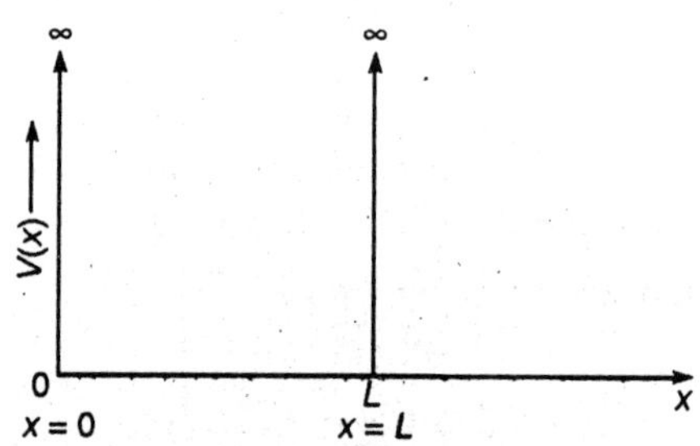


**Figure 3.1** One-dimensional potential box bounded by infinite potential energy barrier

In the present case, this problem is reduced to single electron problem and assumed that (1) the valence electrons move independently of one another in the potential of the metal ions and other electrons, (2) the potential in the interior of the metal is constant, (3) the electron energy distribution is given by the Fermi-Dirac distribution.

Consider a free electron gas in one-dimensional crystal of length $L$ and assume that the potential energy is constant everywhere within this crystal and is equal to zero and at the ends of the crystal, electrons are prevented from leaving the crystal by an infinite potential energy barrier. This means

$$\left.\begin{aligned} V(x) &= 0 \qquad && 0 < x \leq L \\ &= \infty && x < 0 \\ & && x > L \end{aligned}\right\} \tag{3.12}$$

If $\psi$ be the wave function of the electron, the time independent *Schrödinger equation* in one dimension is

$$\frac{d^2\psi}{dx^2}+\frac{2m}{\hbar^2}(E-V)\psi=0, \tag{3.13}$$

where $m$ is the rest mass of the electron, $E$ is the total energy and $V$ is its potential energy. Since potential energy $V$ is zero inside the crystal, then the equation (3.13) becomes

$$\frac{d^2\psi}{dx^2}+\frac{2m}{\hbar^2}E\psi=0. \tag{3.14}$$

But $E=\frac{p^2}{2m}$, where $p$ is the momentum of the electron.

$$\therefore \quad \frac{2m}{\hbar^2}E=\frac{2m}{\hbar^2}\frac{p^2}{2m}$$

$$=\frac{\hbar^2K^2}{\hbar^2} \qquad \text{as } p=\hbar K$$

$$=K^2. \tag{3.15}$$

$$\therefore \quad K=\left(\frac{2m}{\hbar^2}E\right)^{1/2}. \tag{3.16}$$

The equation (3.14) can be written with the help of equation (3.15) as

$$\frac{d^2\psi}{dx^2}+K^2\psi=0. \tag{3.17}$$

The general solution of equation (3.17) is

$$\psi = A\sin Kx + B\cos Kx. \tag{3.18}$$

The boundary conditions of the wave function $\psi$ are

$$\psi(0)=0$$

$$\psi(L)=0. \tag{3.19}$$

Using the boundary conditions (3.19), the equation (3.18) becomes

$$\psi = A\sin Kx \tag{3.20}$$

and $\sin KL = 0$

or, $KL = n\pi$, where $n = 1, 2, 3, \ldots$

$$\therefore \quad K=\frac{n\pi}{L}. \tag{3.21}$$

$n = 0$ is not possible because if $n = 0$, $K = 0$, i.e., $p = 0$, but the momentum of electron $p=\hbar K$ cannot be zero. Hence, the equation (3.20) becomes

$$\psi_n = A\sin\frac{n\pi x}{L}. \tag{3.22}$$

$$\therefore \qquad \frac{d\psi_n}{dx} = \frac{n\pi}{L} A \cos \frac{n\pi x}{L}$$

and
$$\frac{d^2\psi_n}{dx^2} = -\frac{n^2\pi^2}{L^2} A \sin \frac{n\pi x}{L}. \tag{3.23}$$

Substituting the value of $\frac{d^2\psi_n}{dx^2}$ from equation (3.23) in equation (3.17), we have

$$\frac{n^2\pi^2}{L^2} A \sin \frac{n\pi x}{L} = K^2 A \sin \frac{n\pi x}{L}$$

or,
$$K^2 = \frac{n^2\pi^2}{L^2}. \tag{3.24}$$

From equation (3.24) the energy eigenvalue $E_n$ is given by

$$E_n = \frac{\hbar^2 K^2}{2m} = \frac{\hbar^2}{2m}\frac{n^2\pi^2}{L^2}. \tag{3.25}$$

For one dimension and the energy being a quadratic function of the quantum number $n$, the eigenvalues of energy $E_n$ of the electron are:

$$E_1 = \frac{\hbar^2\pi^2}{2mL^2},\ E_2 = \frac{4\hbar^2\pi^2}{2mL^2},\ E_3 = \frac{9\hbar^2\pi^2}{2mL^2}$$

and so on.

To evaluate the constant $A$ of equation (3.22) we choose unit probability of finding the electron somewhere on the line. Using normalisation condition the wave function must satisfy the condition

$$\int_0^L \psi_n^*(x)\, \psi_n(x)\, dx = 1 \tag{3.26}$$

or,
$$\int_0^L A^2 \sin^2\left(\frac{n\pi x}{L}\right) dx = 1$$

or,
$$\frac{A^2}{2}\int_0^L \left[1 - \cos\left(\frac{2n\pi x}{L}\right)\right] dx = 1$$

or,
$$\frac{A^2}{2} L = 1$$

or,
$$A = \left(\frac{2}{L}\right)^{1/2}.$$

$\therefore$ the equation (3.22) becomes

$$\psi_n(x) = \left(\frac{2}{L}\right)^{1/2} \sin \frac{n\pi x}{L}. \tag{3.27}$$

Now suppose we accommodate $N$ electrons on the line $L$. According to Pauli's exclusion principle no two electrons can have all the quantum numbers identical, i.e., each quantum state can be occupied by at most one electron. In a solid the quantum numbers of an electron state are $n$ and $m_s$, where $n$ is the *principal quantum number* and $m_s$ spin quantum number. $n$ is any positive integer and has values 1, 2, 3 ... etc. and spin quantum number $m_s$ has two values $+\frac{1}{2}$ and $-\frac{1}{2}$. Each energy levelled by the quantum number $n$ can accommodate maximum two electrons.

The *Fermi energy* is the energy corresponding to the topmost filled level. Let $n_F$ be the topmost filled level and $n_F$ represent the principal quantum number of the *Fermi level.* If $N$ be the total number of filled electrons to be accomodated then for $N$, an even number,

$$2n_F = N. \tag{3.28}$$

The Fermi energy $E_F$ is given by equation (3.25) as

$$E_F = \frac{\hbar^2}{2m}\left(\frac{n_F \pi}{L}\right)^2, \tag{3.29}$$

where $n = n_F$.

With the help of equation (3.28) the equation (3.29) becomes

$$E_F = \frac{\hbar^2}{2m}\left(\frac{N\pi}{2L}\right)^2, \tag{3.30}$$

i.e., individual energy between $n = 1$ and $n_F = \frac{1}{2}N$, where $\frac{1}{2}$ arises from spin degeneracy.

The total energy $E_0$ of $N$ electrons in the lowest energy state, i.e., the ground state of the entire system is given by

$$E_0 = 2\sum_{n=1}^{N/2} E_n = 2\frac{\hbar^2}{2m}\left(\frac{\pi}{L}\right)^2 \sum_{n=1}^{N/2} n^2$$

$$\cong 2\frac{\hbar^2}{2m}\left(\frac{\pi}{L}\right)^2 \frac{1}{3}\left(\frac{N}{2}\right)^3. \tag{3.31}$$

$$\left[\sum_{n=1} s^2 = \frac{s}{6}(2s+1)(s+1) = \frac{1}{6}s\left(2s^2+3s+1\right) \cong \frac{1}{3}s^3 \ \textit{for } s \gg 1\right]$$

Total ground state energy of $N$ electrons is given by equation (3.31) as

$$E_0 = 2\frac{\hbar^2}{2m}\left(\frac{\pi}{L}\right)^2 \frac{1}{3}\left(\frac{N}{2}\right)^3. \tag{3.32}$$

The equation (3.32) with the help of equation (3.30) takes the form:

$$E_0 = \frac{1}{3} N E_F. \tag{3.33}$$

Thus the average kinetic energy $\left(\frac{E_0}{N}\right)$ in the ground state is approximately one-third of the *Fermi energy*.

*Density of states*: Density of states $D(E)$ is defined as the number of electronic states per unit energy range. From equation (3.25), we know

$$E_n = \frac{\hbar^2}{2m}\frac{\pi^2}{L^2}n^2.$$

$$\therefore \qquad \frac{dE_n}{dn} = \frac{\hbar^2}{m}\frac{\pi^2}{L^2}n$$

or,
$$\frac{dn}{dE_n} = \frac{mL^2}{\hbar^2\pi^2}\frac{1}{n}. \qquad (3.34)$$

$\frac{dn}{dE_n}$ gives the number of energy level per unit energy. As there are two quantum states corresponding to each energy level, one with spin up (↑) and the other with spin down (↓), so the density of states of a free electron gas in one dimension is

$$D(E_n) = 2\frac{dn}{dE_n} = \frac{2mL^2}{\hbar^2\pi^2}\frac{1}{n} = \frac{8mL^2}{h^2}.\frac{1}{n}. \qquad (3.35)$$

Again, the equation (3.25) can be written as

$$E_n = \frac{\hbar^2}{2m}.\frac{\pi^2}{L^2}n^2 = \frac{h^2}{8mL^2}n^2$$

or,
$$n^2 = \frac{8mL^2}{h^2}E_n$$

or,
$$n = \sqrt{8mE_n}\,\frac{L}{h}$$

or,
$$\frac{1}{n} = \left(\frac{h}{L}\right)\frac{1}{\sqrt{8mE_n}}. \qquad (3.36)$$

Using equation (3.36), the equation (3.35) becomes

$$D(E_n) = \frac{4L}{h}\left(\frac{m}{2E_n}\right)^{1/2}. \qquad (3.37)$$

The variation of density of states $D(E_n)$ with energy $E_n$ is plotted in Figure 3.2. Figure shows that the levels are filled from $E_n = 0$ to $E_n = E_F$. All the states are empty beyond $E_F$.

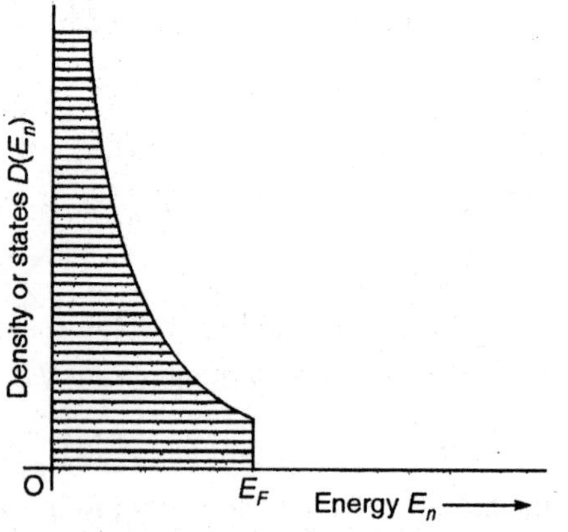


**Figure 3.2** Variation of density or states with energy at 0 °K

## 3.4 FREE ELECTRON THEORY OF SOMMERFELD IN THREE DIMENSIONS

The motion of the electrons is considered as the propagation of the particle wave which for the stationary state is determind by the time-independent Schrödinger equation in three dimensions as

$$\nabla^2\psi(\vec{r}) + \frac{2m}{\hbar^2}(E - V)\,\psi(\vec{r}) = 0\,, \tag{3.38}$$

where $\psi(\vec{r})$ is the wave function, $E$ is the total energy of the electron, $V$ is its potential energy and $m$ is the rest mass of the electron.

If the particle is confined in a cube of edge $L$, the potential $V = 0$ inside the box, then the equation (3.38) takes the form:

$$\nabla^2\psi(\vec{r}) + \frac{2m}{\hbar^2}E\,\psi(\vec{r}) = 0\,. \tag{3.39}$$

To describe the electron within the lattice we apply the periodic boundary condition, i.e., we require the wavefunction to be periodic in $x$, $y$, $z$ with period $L$. The boundary conditions are:

$$\psi(\vec{r}) = \psi(x, y, z) = \psi(x + L, y, z) = \psi(x, y + L, z) = \psi(x, y, z + L)\,. \tag{3.40}$$

By virtue of de Broglie relation, we know,

$$\vec{p} = \hbar\vec{K}\,, \tag{3.41}$$

where $\vec{p}$ is the momentum of the electron and $K$ is the wave number and $K = \frac{2\pi}{\lambda}$, where $\lambda$ is the wavelength of the electrons of momentum $\vec{p}$.

The energy of the electron is given by the relation:

$$E = \frac{p^2}{2m} = \frac{\hbar^2}{2m}K^2 = \frac{\hbar^2}{2m}(K_x^2 + K_y^2 + K_z^2)\,, \tag{3.42}$$

where $K_x$, $K_y$ and $K_z$ are the $x$-component of the wave vector, $y$-component of the wave vector and $z$-component of wave vector respectively.

From equation (3.42), we can write,

$$K^2 = \frac{2mE}{\hbar^2}\,. \tag{3.43}$$

The equation (3.39) with the help of equation (3.43) becomes

$$\nabla^2\psi(\vec{r}) + K^2\psi(\vec{r}) = 0\,. \tag{3.44}$$

The wave function satisfying the free particle Schrödinger equation (3.44) is like plane travelling wave and is of the type:

$$\psi(\vec{r}) = c\,e^{i\vec{K}.\vec{r}} = c\,e^{i(K_x x + K_y y + K_z z)} \tag{3.45}$$

and has the normalization condition:

$$\int_0^V \psi^*(\vec{r})\,\psi(\vec{r})\,d\vec{r}^3 = 1 \tag{3.46}$$

or,

$$\int_0^L\int_0^L\int_0^L c^2\,e^{-i(\vec{K}.\vec{r})}\,e^{i(\vec{K}.\vec{r})}\,dxdydz = 1 \tag{3.47}$$

or, $$c^2 \int_0^L \int_0^L \int_0^L dxdydz = 1$$

or, $$c^2 = \frac{1}{L^3}$$

or, $$c = \frac{1}{L^{3/2}} = \left(\frac{1}{V}\right)^{1/2}. \tag{3.48}$$

The equation (3.45) with the help of equation (3.48) becomes

$$\psi(\vec{r}) = \left(\frac{1}{V}\right)^{1/2} e^{i(K_x x + K_y y + K_z z)}. \tag{3.49}$$

Applying the periodicity conditions to the wave function in (3.49) yield

$$e^{i[K_x(x+L)+K_y y+K_z z]} = e^{i(K_x x+K_y y+K_z z)}$$

or, $$e^{iK_x L} = 1 \tag{3.50}$$

or, $$K_x = 0, \pm\frac{2\pi}{L}, \pm\frac{4\pi}{L}, \ldots, \pm\frac{2m\pi}{L}$$

and similarly, we can get $K_y$ and $K_z$. Thus any component of $\vec{K}$ is of the form $\frac{2n\pi}{L}$, where $n$ is a positive or negative integer. The equation (3.49) can be written as

$$\psi(\vec{r}) = \left(\frac{1}{L^3}\right)^{\frac{1}{2}} e^{[i\frac{2\pi}{L}(K_x x+K_y y+K_z z)]}. \tag{3.51}$$

In the ground state of a system of $N$ free electrons the occupied states may be represented as the points inside a sphere in $K$-space. The energy at the surface of the sphere, i.e., the Fermi energy, ($E_F$, the wave vectors at the Fermi surface) has a magnitude $K_F$ such that

$$E_F = \frac{\hbar^2}{2m} K_F^2. \tag{3.52}$$

The Figure 3.3 represents the ground state of a system of $N$ free electrons, the occupied state of the system fills a sphere of radius $K_F$.

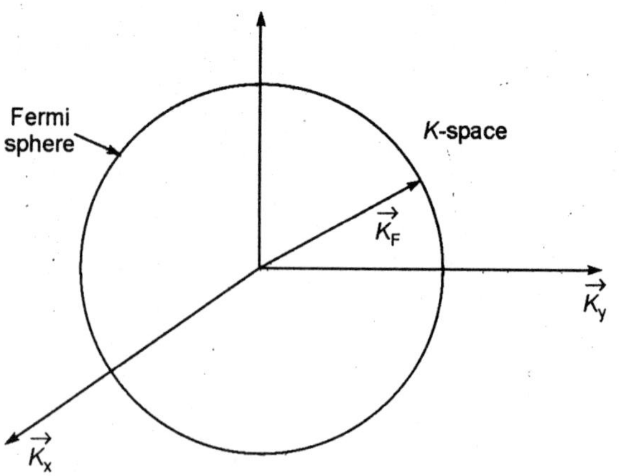


**Figure 3.3** A sphere of radius $K_F$ in $K$-space

Thus the sphere has a volume $\frac{4}{3}\pi K_F^3$ and the total number of allowed electronic states is

$$2\frac{\frac{4}{3}\pi K_F^3}{\left(\frac{2\pi}{L}\right)^3} = \frac{L^3}{3\pi^2} K_F^3 = N, \tag{3.53}$$

where the factor 2 on the left-hand side comes from the two allowed values of spin quantum number $m_s = \pm\frac{1}{2}$ for each allowed value of $\vec{K}$.

From equation (3.53), we get

$$K_F = \left(\frac{3\pi^2 N}{L^3}\right)^{1/3} = \left(\frac{3\pi^2 N}{V}\right)^{1/3}. \tag{3.54}$$

From equation (3.52) with the help of equation (3.54), we obtain

$$E_F = \frac{\hbar^2}{2m}\left(\frac{3\pi^2 N}{V}\right)^{2/3}. \tag{3.55}$$

The equation (3.55) shows the relation between the Fermi energy to the electron concentration $\left(\frac{N}{V}\right)$ and the mass of the electron.

The electron velocity $v_F$ at the Fermi surface is obtained with the help of equations (3.41) and (3.53) as

$$v_F = \frac{\hbar K_F}{m} = \frac{h}{m}\left(\frac{3\pi^2 N}{V}\right)^{1/3}. \tag{3.56}$$

**Density of states**

Density of states $D(E)$ is defined as the number of electronic states per unit energy range, i.e.,

$$D(E) = \frac{dN}{dE}. \tag{3.57}$$

We know from equation (3.55),

$$E_F = \frac{\hbar^2}{2m}\left(\frac{3\pi^2 N}{V}\right)^{2/3}.$$

Hence,

$$N = \frac{V}{3\pi^2}\left(\frac{2m}{\hbar^2}\right)^{3/2} E_F^{3/2}. \tag{3.58}$$

As all the levels below Fermi level are filled, we have

$$N = \frac{V}{3\pi^2}\left(\frac{2m}{\hbar^2}\right)^{3/2} E^{3/2} \tag{3.59}$$

where $E$ refers to as the highest filled energy level.

$$D(E) = \frac{dN}{dE} = \frac{V}{2\pi^2}\left(\frac{2m}{\hbar^2}\right)^{3/2} E^{1/2}. \tag{3.60}$$

The equation (3.60) is true at 0 °K. The density of filled electronic states at a temperature $T$ K is given by

$$N(E) = D(E)\ F(E), \tag{3.61}$$

where $F(E)$ is the Fermi-Dirac function and $F(E)$ is given by

$$F(E) = \frac{1}{e^{(E-E_F)/KT}+1}, \tag{3.62}$$

where $E_F$ is the Fermi energy and $K$ is the Boltzmann's constant. The energy distribution, that is the number $N(E)\ dE$ of electrons with energies between $E$ and $E + dE$ at a temperature $T$, is given by

$$N(E)\ dE = D(E)\ F(E)\ dE$$

$$= \frac{V}{2\pi^2}\left(\frac{2m}{\hbar^2}\right)^{3/2} \frac{E^{1/2}\ dE}{e^{(E-E_F)/KT}+1}. \tag{3.63}$$

The function $F(E)$ assumes only values between 0 and 1, i.e.,

$$0 \le F(E) \le 1.$$

In particular, we have

$$\begin{array}{lll} T = 0\ °K & E > E_{F_0} & F(E) = 0 \\ & E \le E_{F_0} & F(E) = 1 \\ T > 0\ °K & E >> E_F & F(E) \simeq 0 \\ & E << E_F & F(E) \simeq 1 \end{array} \tag{3.64}$$

$$E = E_F \quad F(E) = \frac{1}{2}. \tag{3.65}$$

For energies above $E_F$ such that $(E - E_F) > KT$, the term 1 in the denominator of equation (3.63) may be neglected as compared to the other terms, so the equation (3.63) becomes

$$N(E)dE = \frac{V}{2\pi^2}\left(\frac{2m}{\hbar^2}\right)^{3/2} \frac{E^{1/2} dE}{e^{(E-E_F)/KT}} \tag{3.66}$$

*The average Kinetic Energy of an electron at absolute zero*:

The average kinetic energy of an electron at absolute zero is given by

$$\overline{E}_0 = \frac{1}{N}\int_0^\infty E\ D(E)\ F(E)\ dE \tag{3.67}$$

Using the equation (3.60) the equation (3.67) becomes

$$\overline{E}_0 = \frac{1}{N}\int_0^\infty E \frac{V}{2\pi^2}\left(\frac{2m}{\hbar^2}\right)^{3/2} E^{1/2}\ F(E)\ dE$$

$$= \frac{V}{2\pi^2}\left(\frac{2m}{\hbar^2}\right)^{3/2} \frac{1}{N}\left[\int_0^{E_{F_0}} E^{3/2}\ F(E)\ dE + \int_{E_{F_0}}^\infty E^{3/2}\ F(E)\ dE\right]. \tag{3.68}$$

Using the equation (3.64), the equation (3.68) becomes

$$\vec{E}_0 = \frac{V}{2\pi^2}\left(\frac{2m}{\hbar^2}\right)^{3/2} \frac{1}{N} \int_0^{E_{F0}} E^{3/2}\, dE .$$

$$\therefore \qquad \vec{E}_0 = \frac{2}{5}\frac{V}{2\pi^2}\left(\frac{2m}{\hbar^2}\right)^{3/2} \frac{1}{N} E_{F_0}^{5/2} . \qquad (3.69)$$

With the help of equation (3.59) the equation (3.69) becomes

$$\vec{E}_0 = \frac{3}{5} E_{F_0} . \qquad (3.70)$$

The equation (3.70) gives the average kinetic energy at absolute zero to the Fermi energy at absolute zero.

## SOLVED PROBLEMS

1. Calculate the Fermi energy and Fermi velocity of the electrons at 0 °K for a metal with electron density of $10^{28}$ m$^{-3}$.

**Soln.** The Fermi energy at 0 °K is given by the equation (3.55) as

$$E_{F_0} = \frac{\hbar^2}{2m}\left(3\pi^2 \frac{N}{V}\right)^{2/3} .$$

Electron density $\frac{N}{V}$ = $10^{28}$ m$^{-3}$ (given).

We know, $\hbar = 1.056 \times 10^{-34}$ Js

Rest mass of the electron, $m = 9.11 \times 10^{-31}$ kg.

$$\therefore \qquad E_{F_0} = \frac{\left(1.056 \times 10^{-34}\right)^2}{2 \times 9.11 \times 10^{-31}} \times \left(3\pi^2 \times 10^{28}\right)^{2/3} \text{ J}$$

$$= 2.6896 \times 10^{-18} \text{ J}$$

$$= \frac{2.6896 \times 10^{-18}}{1.6 \times 10^{-19}} \text{ eV}$$

$$= 16.81 \text{ eV.}$$

The Fermi velocity $v_F$ at absolute zero is given by

$$v_{F_0} = \frac{\hbar}{m}\left(3\pi^2 \frac{N}{V}\right)^{1/3} = \frac{1.056 \times 10^{-34}}{9.11 \times 10^{-31}}\left(3\pi^2 \times 10^{28}\right)^{1/3}$$

$$= \mathbf{7.7 \times 10^{6}\ ms^{-1}}.$$

2. Consider silver in the metallic state with one free electron per atom. Calculate the Fermi energy of silver. Given the density of silver is 10.5 gm-cm$^{-3}$, atomic weight 108. Avogadro number $N = 6.02 \times 10^{23}$ **[H.P.U. 1996]**

**Soln.** Volume of 1 gm-mole of silver $V = \dfrac{\text{Molecular Weight } (M)}{\text{Density } (\rho)}$

$$= \frac{108}{10.5}\ \text{cm}^3$$

$$= 10.285\ \text{cm}^3.$$

Electron density, $\dfrac{N}{V} = \dfrac{6.02 \times 10^{23}}{10.285} \times 1$ free electrons per cc

$$= 5.853 \times 10^{28} \text{ free electrons per m}^3.$$

Fermi energy, $E_F = \dfrac{\hbar^2}{2m}\left(3\pi^2 \dfrac{N}{V}\right)^{2/3}$ J

$$= \frac{\left(1.056 \times 10^{-34}\right)^2}{2 \times 9.11 \times 10^{-31}}\left(3\pi^2 \times 5.853 \times 10^{28}\right)^{2/3} \text{ J}$$

$$= 8.83 \times 10^{-19} \text{ J}$$

$$= \mathbf{5.518\ eV.}$$

3. The atomic radius of sodium is 1.85 Å. Calculate the Fermi energy of sodium at absolute zero.

**Soln.** Sodium is a BCC structure.
Radius of the sodium atom is given as $r = 1.85$Å
$= 1.85 \times 10^{-10}$ m.
But we know from Art. 1.5(iii), lattice distance

$$a = \frac{4r}{\sqrt{3}}.$$

$\therefore$ volume of the unit cell, $V = a^3$

$$= \left(\frac{4 \times 1.85 \times 10^{-10}}{\sqrt{3}}\right)^3$$

$$= 7.7985 \times 10^{-29} \text{ m}^3.$$

Since sodium atom has only one valence electron the number of free electrons in a unit cell of sodium is two.

$$\therefore \quad \frac{N}{V} = \frac{2}{7.7985 \times 10^{-29}} = 2.564 \times 10^{28} \text{ electrons/m}^3.$$

So the Fermi energy of sodium is given by the equation (3.55) as

$$E_{F_0} = \frac{\hbar^2}{2m}\left(3\pi^2 \frac{N}{V}\right)^{2/3}.$$

But we know, $\hbar = 1.056 \times 10^{-34}$ Js

$m = 9.11 \times 10^{-31}$ kg.

$$\therefore \quad E_{F_0} = \frac{\left(1.056\times10^{-34}\right)^2}{2\times9.11\times10^{-31}}.\left(3\pi^2 2.564\times10^{28}\right)^{2/3} \text{ J}$$

$= \mathbf{2.9\ eV}$.

## REVIEW QUESTIONS

1. State and establish Wiedemann-Franz law.
2. Describe the free electron gas model of metals.
3. What is the Fermi energy? Derive an expression for the Fermi energy of a free electron metal at absolute zero temperature.
4. Show that the average kinetic energy per electron for a three-dimensional free electron gas at 0 °K is $\bar{E}_0 = \left(\frac{3}{5}\right) E_{F_0}$, where $E_{F_0}$ is the Fermi Energy at 0 °K.
5. Explain the concept of density of states for a free elctron gas.
6. What is Fermi level? Derive an expression for the Fermi energy in terms of the number of electrons per unit volume.
7. Show that for a one-dimensional metallic crystal the average kinetic energy of electron in the ground state is one-third of the Fermi energy.

## PROBLEMS

1. The density of potassium is 860 kg m$^{-3}$. Calculate the Fermi energy of potassium at 0 °K. Atomic mass of potassium = 39.1 amu. **[Ans.** 2.02 eV]
2. Evaluate the Fermi energy of the metal lithium from the following data:
   Density of lithium = 534 kg/m$^3$, Atomic weight of lithium = $6.939 \times 10^{-3}$ kg/m$^3$,
   $\hbar = 1.054 \times 10^{-34}$ joules-sec,
   Mass of the electron = $9.1 \times 10^{-31}$ kg. **[Ans.** 4.70 eV]
3. There are $2.5 \times 10^{28}$ free electrons per cubic metre of sodium. Calculate the Fermi energy and Fermi velocity. **[Bang. U. 1995]**
   **[Ans.** $E_F = 3.1$ eV $\quad v_F = 1.048 \times 10^6$ m/sec]
4. Aluminium metal has FCC structure and each atom contributes single electron as free elctron and its lattice constant is 4 Å. Calculate (i) Fermi energy ($E_F$) (ii) Total kinetic energy of free electron gas per unit volume at 0 °K.
   Given: $\hbar = 1.054 \times 10^{-34}$ Js.
   Rest mass of electron = $9.11 \times 10^{-31}$ kg.
   **[Ans.** (i) 2.28 eV (ii) $2.14 \times 10^{28}$ eV per m$^3$]

# 4
# *Band Theory of Solids*

## 4.0 INTRODUCTION

The free electron theory was successful in explaining various properties of metals but there are various other properties which could not be explained by this theory. The quantum free electron theory of metals assumes that a conduction electron in a metal experiences a constant or zero potential and so is completely free to move about the crystal which is restrained only by the surface of the crystal. Within the framework of one electron theory there are two different approaches:

1. *Heitler-London valence band* scheme which is applicable when the atoms are far apart.
2. Another approach is that of Bloch which is considered that an electron belongs to the crystal as a whole rather than to a particular atom.

The band structure of a solid can be described by the nearly free electron model in which the band of electrons in a solid crystal is supposed to move in a region of periodically varying potential due to ion cores of the solid crystals. In this model we assume the following:

1. The atomic nuclei at their lattice sites are perfectly at rest. Interactions between electrons and phonons are neglected.
2. The interaction between individual electrons are neglected. Every electron is in the potential field which is due to the positive nuclei and an average charge distribution of other electrons which may be taken constant.

The periodic potential extends up to infinity in all directions except at the surface of the crystal where periodicity is interrupted and this type of irregularity may be ignored.

The motion of an electron in a periodic potential has two characteristc features:

1. There exist allowed frequency bands separated by forbidden region or band gaps.
2. The frequency is no longer proportional to wave number $\vec{K}$ but periodic function of wave number.

In chapter 3 in equation (3.25) we see that the motion of the electrons in a constant potential in free electron theory is given by

$$E = \frac{\hbar^2 K^2}{2m},$$

where $E$ is the energy and $K$ is the wave number and $K = \frac{2\pi}{\lambda} = \frac{p}{\hbar}$, $\lambda$ is the wavelength associated with the electron and $p$ is the momentum of the electron and potential energy is taken zero. If we consider the motion of an electron in periodic potential it will be shown that (1) there exist allowed energy bands separated by forbidden region, (2) the function $E(\vec{K})$ is periodic in $\vec{K}$.

One-dimensional representation of periodic potential experienced by an electron in a periodic crystal lattice is shown in Figure 4.1.

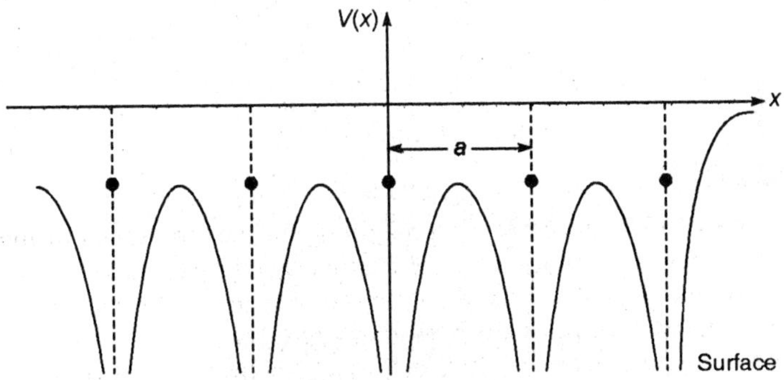


**Figure 4.1** Representation of one-dimensional potential experienced by an electron in a perfectly periodic lattice. Here *a* is the lattice constant

## 4.1 THE BLOCH THEOREM

Consider an electron moving on its own accord inside a crystal; then the Schrödinger equation for an electron moving in one-dimensional periodic potential is

$$H\,\psi(x) = E\,\psi(x), \tag{4.1}$$

where $H$ is the Hamiltonian for the electron and in one dimension.

$$H = -\frac{\hbar^2}{2m}\frac{d^2}{dx^2} + V(x), \tag{4.2}$$

where $V(x)$ is the potential at a distance $x$ from the origin and is assumed to vary periodically in space with an interval equal to the lattice constant. The equation (4.1) can be written with the help of equation (4.2) as

$$\frac{d^2\psi}{dx^2} + \frac{2m}{\hbar^2}[E - V(x)]\psi(x) = 0. \tag{4.3}$$

If the crystal potential has the periodic property with period equal to the lattice constant $a$, we have

$$V(x) = V(x + a). \tag{4.4}$$

The solution of equation (4.3) is expressed by Bloch theorem or Floquent theorem which states that the solution of the theorem exists in the form

$$\psi(x) = \psi_K(x) \simeq u_K(x)\, e^{\pm ikx}, \tag{4.5}$$

where $u_K(x)$ has the same periodicity as the crystal potential, i.e.,

$$u_K(x) = u_K(x + a). \tag{4.6}$$

Bloch function $\psi(x) = e^{ikx}\, u_K(x)$ has the property

$$\psi(x+a) = e^{ik(x+a)}\, u_K(x+a)$$

$$= e^{ikx}\, u_K(x)\, e^{ika}$$

$$\psi(x)\, e^{ika} = \lambda\psi(x), \tag{4.7}$$

since $u_K(x + a) = u_K(x)$ and $\lambda = e^{ika}$.

Suppose, $f(x)$ and $g(x)$ are two independent solutions of the Schrödinger equation (4.3); the solution of equation (4.3) can be written as

$$\psi(x) = Af(x) + B\, g(x), \tag{4.8}$$

where $A$ and $B$ are two arbitrary constants. Since the potential in equation (4.4) is periodic, then the functions $f(x + a)$ and $g(x + a)$ are the solution of Schrödinger equation (4.3), we have the relation

$$f(x + a) = c_1 f(x) + c_2 g(x) \tag{4.9}$$

$$g(x + a) = d_1 f(x) + d_2 g(x), \tag{4.10}$$

where $c_1$, $c_2$, $d_1$ and $d_2$ are the real functions of energy $E$.

The general solution of equation (4.3) with the help of equations (4.9) and (4.10) is

$$\psi(x + a) = Af(x + a) + Bg(x + a). \tag{4.11}$$

With the help of equation (4.7) the equation (4.11) becomes

$$\psi(x+a) = A[c_1 f(x)+c_2 g(x)] + B[d_1 f(x) + d_2 g(x)]$$

$$= (Ac_1 + Bd_1)\, f(x) + (Ac_2 + Bd_2)\, g(x). \tag{4.12}$$

Choose the constants $A$ and $B$ such that

$$Ac_1 + Bd_1 = \lambda A \tag{4.13}$$

$$Ac_2 + Bd_2 = \lambda B, \tag{4.14}$$

where $\lambda$ is a constant.

Thus we obtain the property

$$\psi(x + a) = \lambda\psi(x). \tag{4.15}$$

Since the equations (4.13) and (4.14) have nonvanishing solution for $A$ and $B$, only if the determinant of their coefficient vanishes, i.e.,

$$\begin{vmatrix} c_1 - \lambda & d_1 \\ c_2 & d_2 - \lambda \end{vmatrix} = 0 \tag{4.16}$$

Expanding and rearranging the determinant in equation (4.16), we get

$$\lambda^2 - \lambda(c_1 + d_2) + (c_1 d_2 - c_2 d_1) = 0 \tag{4.17}$$

Now we shall first prove that $(c_1d_2 - c_2d_1) = 1$ in the following manners: as $f(x)$ and $g(x)$ are two independent real solutions of Schrödinger's equation, then we have from equation (4.3),

$$\frac{d^2 f(x)}{dx^2} + \frac{2m}{\hbar^2}\left[E - V(x)\right] f(x) = 0 \tag{4.18}$$

and

$$\frac{d^2 g(x)}{dx^2} + \frac{2m}{\hbar^2}\left[E - V(x)\right] g(x) = 0\,. \tag{4.19}$$

Multiplying equation (4.18) by $g(x)$ and equation (4.19) by $f(x)$ respectively and substracting, we get

$$g(x)\frac{d^2 f(x)}{dx^2} - f(x)\frac{d^2 g(x)}{dx^2} = 0\,. \tag{4.20}$$

Integrating equation (4.20), we get

$$g(x)\frac{df(x)}{dx} - f(x)\frac{dg(x)}{dx} = \text{constant} = W(x). \tag{4.21}$$

$W(x)$ is called *Wronskian of the two solutions* and is constant. It must have the same value at $x$ and $x + a$, i.e.,

$$W(x) = W(x + a) \tag{4.22}$$

$$W(x+a) = g(x+a)\frac{d}{dx} f(x+a) - f(x+a)\frac{d}{dx} g(x+a)\,. \tag{4.23}$$

Substituting the value of $f(x + a)$ and $g(x + a)$ from equations (4.9) and (4.10), we get from equation (4.23)

$$W(x + a) = (c_1d_2 - c_2d_1)\, W(x). \tag{4.24}$$

Since $W(x + a) = W(x)$, we get $c_1d_2 - c_2d_1 = 1$.

Therefore, the equation (4.17) becomes

$$\lambda^2 - \lambda(c_1 + d_2) + 1 = 0\,. \tag{4.25}$$

The quantity $(c_1 + d_2)$ is a real function of energy $E$. Let the two roots of $\lambda$ in equation (4.25) be $\lambda_1$ and $\lambda_2$. Therefore, from equation (4.25) we get $\lambda_1\lambda_2 = 1$.

The two functions $\psi_1(x)$ and $\psi_2(x)$ exhibit the following property:

$$\psi_1(x + a) = \lambda_1 \psi_1(x) \tag{4.26}$$

$$\psi_2(x + a) = \lambda_2 \psi_2(x). \tag{4.27}$$

We consider the two cases of special interest:

**Case I:** For energy range $(c_1 + d_2)^2 < 4$, the two roots of $\lambda$ in equation (4.25) are complex and since $\lambda_1\lambda_2 = 1$, the roots are complex conjngates of each other. Then we may write,

$$\lambda_1 = e^{ika} \tag{4.28}$$

and

$$\lambda_2 = e^{-ika}, \tag{4.29}$$

where $k$ is real.

The corresponding function $\psi_1(x)$ and $\psi_2(x)$ mast satisfy the relation:

$$\psi_1(x + a) = e^{ika}\, \psi_1(x) \tag{4.30}$$

$$\psi_2(x + a) = e^{-ika}\, \psi_2(x). \tag{4.31}$$

The equations (4.30) and (4.31) can be written together as

$$\psi(x + a) = e^{\pm ika} \psi(x). \tag{4.32}$$

The equation (4.32) has the property of Bloch function as given by equation (4.5) and has the form:

$$\psi(x) = e^{\pm ikx} u_K(x), \tag{4.33}$$

where $u_K(x) = u_K(x + a)$ [Equation (4.6)]

To verify the above equation we put $x + a$ for $x$ and get

$$\psi(x + a) = e^{ik(x+a)} u_K(x + a) = e^{\pm ika} e^{\pm ikx} u_K(x)$$

$$= e^{\pm ika} \psi(x) = \lambda\psi(x). \tag{4.34}$$

Therefore, the equations (4.32) and (4.34) are the same. Now the Schrödinger equation

$$\frac{d^2\psi(x)}{dx^2} + \frac{2m}{\hbar^2}\left[E - V(x)\right]\psi(x) = 0$$

has the solution of the property

$$\psi(x + a) = e^{\pm ika} \psi(x).$$

The solution in the form of Bloch function:

$$\psi(x) = e^{\pm ikx} u_K(x)$$

where $u_K(x) = u_K(x + a)$.

**Case II:** For energy range $(c_1 + d_2)^2 > 4$, the roots are real and reciprocal to each other. Two real roots $\lambda_1$ and $\lambda_2$ may be taken as

$$\lambda_1 = e^{\mu a} \quad \text{and} \quad \lambda_2 = e^{-\mu a}.$$

These roots correspond to solution of the Schrodinger equation of the type

$$\psi_1(x) = e^{-\mu x} u(x) \tag{4.35}$$

$$\psi_2(x) = e^{-\mu x} u(x). \tag{4.36}$$

where $\mu$ is a real quantity. Although the above solution is mathematically sound, they cannot be accepted as wave function describing electrons because they are not bounded. Thus, there are electronic states in the energy region corresponding to real roots $\lambda_1$ and $\lambda_2$. Thus, $\lambda_1 = e^{ika}$ and $\lambda_2 = e^{-ika}$ are allowed roots and $\lambda_1 = e^{\mu a}$ and $\lambda_2 = e^{-\mu a}$ are forbidden roots. The allowed roots correspond to the allowed energy region and the forbidden roots are associated with forbidden energy regions. This means that the energy spectrum of an electron in a periodic potential consists of allowed and forbidden regions. This result will be illustrated further in Kronig-Penney model where we consider the motion of electrons in a simple one-dimensional periodic potential.

## 4.2 BOLCH THEOREM IN THREE DIMENSIONS

Consider the motion of an electron moving on its own inside a crystal of three-dimensional lattice; then Schrödinger's equation is

$$\nabla^2\psi(\vec{r}) + \frac{2m}{\hbar^2}\left[E - V(\vec{r})\right]\psi(\vec{r}) = 0, \tag{4.37}$$

where

$$V(\vec{r}) = V\left(\vec{r} + N_1\vec{a}_1 + N_2\vec{a}_2 + N_3\vec{a}_3\right) \tag{4.38}$$

and $\vec{r}$ is the position vector, $\vec{a}_1$, $\vec{a}_2$, $\vec{a}_3$ are translation vectors and $N_1$, $N_2$, $N_3$ are integers.

The solution of equation (4.37) is expressed by Bloch theorem as

$$\psi_K(\vec{r}) = u_K(\vec{r})\, e^{\pm i(\vec{K}\cdot\vec{r})}, \tag{4.39}$$

where

$$u_K(\vec{r}) = u_K(\vec{r} + N_1\vec{a}_1 + N_2\vec{a}_2 + N_3\vec{a}_3) \tag{4.40}$$

and $u_K(\vec{r})$ is periodic.

## 4.3 THE KRONIG-PENNEY MODEL

This model illustrates the behaviour of electrons in periodic potential by assumming simple one-dimensional rectangular square well potential as indicated in Figure 4.2. The period of potential is $(a + b)$. In the region $0 < x < a$ the electron is assumed to be in the vicinity of a nucleus and the potential is taken as zero and in the region $-b < x < 0$ the potential energy is taken as zero ($V_0$). This model is a highly rough approximation of the potential existing inside a lattice. The wave functions associated with this model are obtained by solving Schrödinger equations in two regions as

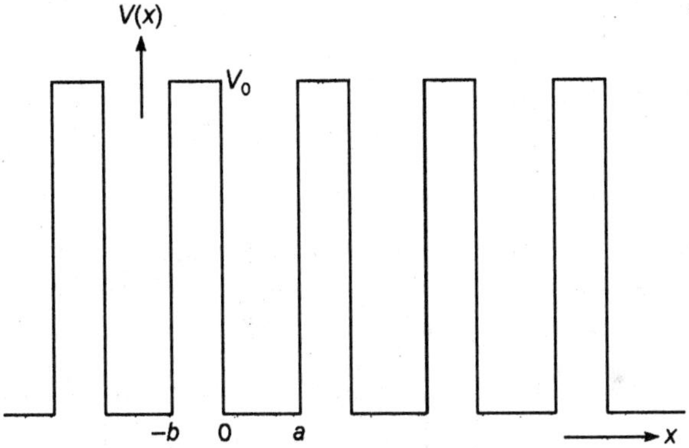


**Figure 4.2** One-dimensional square well potential used by Kronig and Penney

$$\frac{d^2\psi}{dx^2} + \frac{2m}{\hbar^2} E\psi = 0 \quad \text{for} \quad 0 < x < a \tag{4.41}$$

and

$$\frac{d^2\psi}{dx^2} + \frac{2m}{\hbar^2}(E - V_0)\,\psi = 0 \quad \text{for } -b < x < 0. \tag{4.42}$$

Assuming the total energy $E$ of the electron is smaller than $V_0$, we define two real quantities $\alpha$ and $\beta$ such that

$$\alpha^2 = \frac{2m}{\hbar^2} E \tag{4.43}$$

and

$$\beta^2 = \frac{2m}{\hbar^2}(V_0 - E). \tag{4.44}$$

Thus the equations (4.41) and (4.42) become

$$\frac{d^2\psi}{dx^2} + \alpha^2\psi = 0 \tag{4.45}$$

$$\frac{d^2\psi}{dx^2} - \beta^2\psi = 0. \tag{4.46}$$

The solution of the wave equations (4.45) and (4.46) for periodic potential must be of the form of Bloch function and is of the form given by equation (4.5) as

$$\psi(x) = u_K(x)\, e^{ikx}$$

where $u_K(x)$ is the periodic function of $x$ with period $(a + b)$. From equation (4.5), we get

$$\frac{d\psi}{dx} = e^{ikx}\frac{du_K(x)}{dx} + u_K(x)\, ike^{ikx}$$

and

$$\frac{d^2\psi}{dx^2} = e^{ikx}\frac{d^2u_K(x)}{dx^2} + 2ik\, e^{ikx}\frac{du_K(x)}{dx} - k^2e^{ikx}u_K(x)\,. \tag{4.47}$$

Substituting the equation (4.47) in equations (4.45) and (4.46), we get

$$\frac{d^2u_K(x)}{dx^2} + 2iK\frac{du_K(x)}{dx} + (\alpha^2 + K^2)\,u_K(x) = 0 \qquad 0 < x < a \tag{4.48}$$

and

$$\frac{d^2u_K(x)}{dx^2} + 2iK\frac{du_KiK_1(x)}{dx} - (\beta^2 + K^2)\,u_K(x) = 0 \quad -b < x < 0. \tag{4.49}$$

The general solutions of the equations (4.48) and (4.49) are given by

$$u_1(x) = Ae^{i(\alpha-k)x} + Be^{-i(\alpha+k)x} \tag{4.50}$$

$$u_2(x) = Ce^{(\beta-ik)x} + De^{-(\beta+ik)x}, \tag{4.51}$$

where $A$, $B$, $C$, and $D$ are constants. These constants are so chosen that they satisfy the conditions of continuity and periodicity of the wave functions, i.e.,

$$u_1(0) = u_2(0) \tag{4.52i}$$

$$u_1(a) = u_2(-b) \tag{4.52ii}$$

and

$$\left(\frac{du_1(x)}{dx}\right)_{x=0} = \left(\frac{du_2(x)}{dx}\right)_{x=0} \tag{4.53i}$$

$$\left(\frac{du_1(x)}{dx}\right)_{x=a} = \left(\frac{du_2(x)}{dx}\right)_{x=-b}. \tag{4.53ii}$$

Applying the boundary conditions (4.52) and (4.53) from the equations (4.50) and (4.51), we obtain the following equations:

$$A + B = C + D \tag{4.54}$$

$$i(\alpha - k)A - i(\alpha + k)B = (\beta - ik)C - (\beta + ik)D \tag{4.55}$$

$$e^{i(\alpha-k)a}A + Be^{-i(\alpha+k)a}B = e^{-(\beta-ik)b}C + e^{(\beta+ik)b}D \tag{4.56}$$

$$i(\alpha - k)e^{i(\alpha-k)a}A - i(\alpha + k)e^{-i(\alpha+k)a}B$$

$$= (\beta - ik)e^{-(\beta-ik)b} - i(\beta + ik)e^{(\beta+ik)b}D\,. \tag{4.57}$$

The four equations (4.54) to (4.57) will have nonvanishing solutions if the determinant of the coefficients $A$, $B$, $C$ and $D$ vanishes, i.e.,

$$\begin{vmatrix} 1 & 1 & 1 & 1 \\ i(\alpha-k) & -i(\alpha+k) & (\beta-ik) & -(\beta+ik) \\ e^{i(\alpha-k)a} & e^{-i(\alpha+k)a} & e^{(-\beta-ik)a} & e^{i(\beta+ik)a} \\ i(\alpha-k)e^{i(\alpha-k)a} & -i(\alpha+k)e^{i(\alpha+k)a} & (\beta-k)e^{-(\beta-ik)b} & -(\beta+ik)e^{(\beta+ik)b} \end{vmatrix} = 0 .$$

On solving the determinant, we get

$$\frac{\beta^2-\alpha^2}{2\alpha\beta}\sinh\beta b\sin\alpha a+\cosh\beta b\cos\alpha a=\cos k\,(a+b). \qquad (4.58)$$

Equation (4.58) is complicated. To simplify the equation (4.58) Kronig and Penney considered the case such that the potential barriers become delta function, i.e., $V_0 \rightarrow \infty$ and $b \rightarrow 0$ but the product $V_0 b$ or $\beta^2 b$ remains finite. Within this limit, equation (4.58) reduces to

$$\frac{mV_0 b}{\hbar^2\alpha}\sin\alpha a+\cos\alpha a=\cos ka \qquad (4.59)$$

as $b \rightarrow 0$, $\sinh\beta b \rightarrow \beta b$ and $\cosh\beta b = 1$

and
$$\frac{\beta^2-\alpha^2}{2\alpha\beta}=\frac{1}{2\alpha\beta}\left[\frac{2m(V_0-E)}{\hbar^2}-\frac{2mE}{\hbar^2}\right]$$

$$=\frac{2m}{2\alpha\beta\,\hbar^2}(V_0-2E).$$

Since $V_0 >> E$, $\quad\therefore\quad \dfrac{\beta^2-\alpha^2}{2\alpha\beta}=\dfrac{2mV_0}{2\alpha\beta\,\hbar^2}$.

Let us assume $P=\dfrac{2mV_0\,ba}{\hbar^2}$, then the equation (4.59) can be written as

$$P.\frac{\sin\alpha a}{\alpha a}+\cos\alpha a=\cos ka. \qquad (4.60)$$

$P$ is a measure for the area $V_0 b$ of the potential barrier. The physical significance of $P$ is that while attempting increase of $P$, the area of the potential barrier is increased and the given electron is bound more strongly to a particular potenital well. If the left-hand side of equation (4.60) is plotted against $\alpha a$ for $P=\dfrac{3\pi}{2}$, a curve as shown in Figure 4.3 is obtained. It is important to reaslise that the right-hand side of equation (4.61) can accept values between –1 and +1 as indicated in Figure 4.3. Since $\alpha^2$ is proportional to $E$ the abscissa is a measure of energy. The following important conclusions can be drawn from the Figure 4.3.

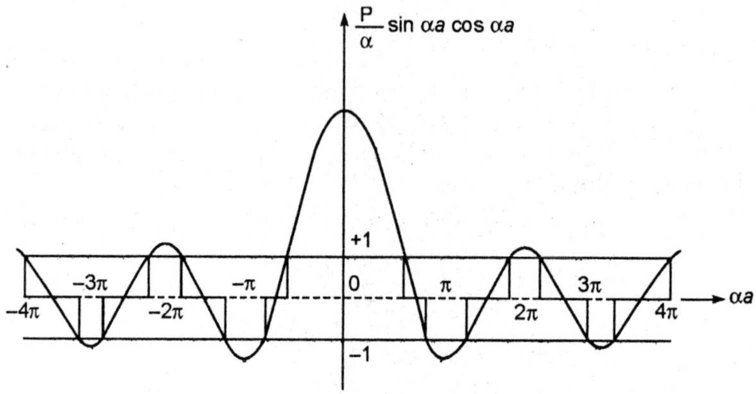


**Figure 4.3** $\left(P\dfrac{\sin\alpha a}{\alpha a}+\cos\alpha a\right)$ is plotted against $\alpha a$ for $P=\dfrac{3\pi}{2}$. The allowed regions are drawn in thick line

(i) The energy spectrum of the electrons consists of alternate region of allowed energy bands and forbidden region.

(ii) The width of the allowed energy bands increases with the increasing energy, i.e., with increasing $\alpha a$.

(iii) The width of a particular allowed band decreases with the increasing value of $P$, i.e., with the increasing binding energy of the electrons. In the limit $P \rightarrow \infty$ the allowed regions become infinitely narrow and spectrum reduces to a single energy level, that is a line spectrum. Accordingly the equation (4.60) has solutions if $\sin \alpha a = 0$.

$$\text{or,} \qquad \alpha a = \pm n\pi \qquad \text{or,} \qquad \alpha^2 = \frac{n^2\pi^2}{a^2},$$

where $n$ is an integer. Therefore, we get from equation (4.43),

$$\alpha^2 = \frac{2mE}{\hbar^2} = \frac{n^2\pi^2}{a^2}$$

$$\text{or,} \qquad E = \left(\frac{\pi^2\hbar^2}{2ma^2}\right)n^2. \tag{4.61}$$

The equation (4.61) gives the energy level of a particle in a constant potential box of atomic dimension. This is expected physically for large $P$ as tunnelling through the barriers becomes impossible. When $P \rightarrow 0$, the potential barriers become very weak, i.e., the electrons are free electrons. In this case, from the equation (4.60), we get

$$\cos \alpha a = \cos ka$$

$$\text{or,} \qquad \alpha = k$$

$$\text{or,} \qquad \alpha^2 = k^2$$

Therefore, from equation (4.61), we get

$$E = \frac{k^2\hbar^2}{2m}. \tag{4.62}$$

This is the same result as obtained by free electron model (equation 3.15).

We summerize these results in Figure 4.4 where energy spectrum is a funtion of $P$. For $P = 0$, we have the free electron model and energy spectrum is quasicontinuous and free; for $P = \infty$, a line spectrum is obtained. For any other value of $P$ the position and width of the allowed and forbidden bands are obtained by erecting a vertical line.

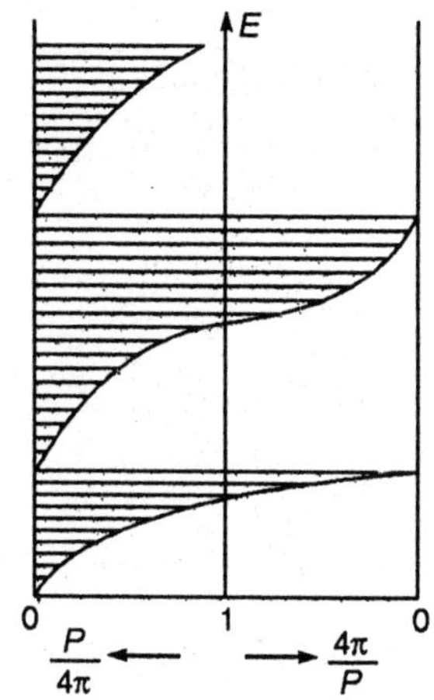


**Figure 4.4** Allowed and forbidden ranges as shown by shaded and open respectively

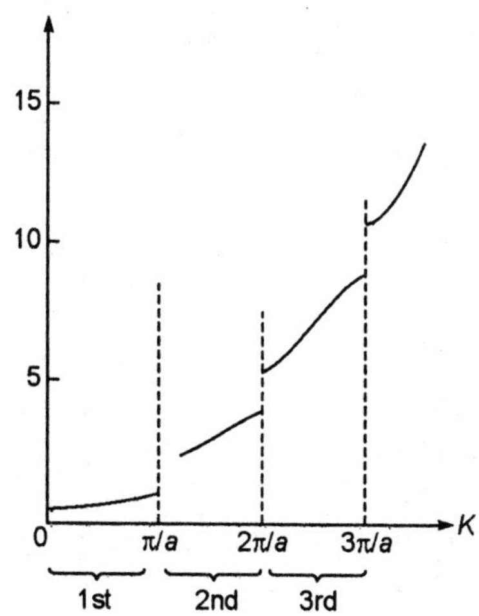


**Figure 4.5** Energy is plotted against $K$ for $P = \frac{3\pi}{2}$

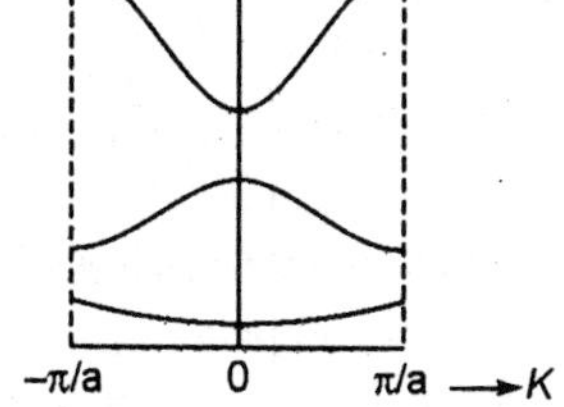


**Figure 4.6** Represents $E$ vs. $K$ curve in zone

From equation (4.62) we obtain the energy $E$ as a function of wave number and the results is represented in Figures 4.5 and 4.6. In the graph $E$ versus $K$ curve the discontinuities occur for $K = \frac{n\pi}{a}$, where $n$ is an integer. The values of $K$ define the first, second, third, etc. *Brillouin zones*. First zone extends from $-\frac{\pi}{a}$ to $\frac{\pi}{a}$, second zone extends from $\frac{\pi}{a}$ to $\frac{2\pi}{a}$ and another part extending between $-\frac{\pi}{a}$ to $-\frac{2\pi}{a}$.

**The number of possible wave number in a band**

We have earlier assumed that the crystal is an infinitely long one-dimensional crystal. Now we investigate the consequences of imposing boundary conditions. For a linear crystal of finite length the boundary condition is

$$\psi(x + \mathrm{L}) = \psi(x). \tag{4.63}$$

Since we are dealing with Bloch functions, this requires

$$e^{ik(x+\mathrm{L})}\, u_K(x + \mathrm{L}) = e^{ikx}\, u_K(x)$$

and because of periodicity of $u_K$ we have

$$u_K(x + \mathrm{L}) = u_K(x).$$

$$\therefore \qquad e^{ik(x+\mathrm{L})} = e^{ikx}$$

$$\text{or,} \qquad e^{ikL} = 1$$

$$\text{or,} \qquad K = \frac{2\pi n}{L}, \text{ where } n = \pm 1, \pm 2, ..., \text{etc.} \tag{4.64}$$

The number of possible wave function in the range $dK$ is

$$dK = \frac{2\pi}{L} dn$$

$$\text{or,} \qquad dn = \frac{L}{2\pi} dK. \tag{4.65}$$

Since $K$ is limited by the relation $K = \frac{n\pi}{a}$, it follows that the maximum value of $n$ is $\frac{L}{2\pi} = \frac{N}{2}$, where $N$ is the number of unit cells. This leads to the condition that the total number of possible wave function in any energy band is equal to the number of unit cell $N$.

## 4.4 THE VELOCITY OF THE ELECTRON ACCORDING TO THE BAND THEORY

According to de Broglie the motion of electron moving with a velocity $v$ may be equivalent to a wave pocket moving with a group velocity $v_g$. Let us consider the velocity of an electron discribed by wave vector $K$. Thus we can write,

$$v = v_g = \frac{d\omega}{dK}, \tag{4.66}$$

where $\omega$ is the angular frequency of the de Broglie waves and it is related to the energy of the particle as

$$E = \hbar\,\omega \tag{4.67}$$

$$\text{or,} \qquad \omega = \frac{E}{\hbar}.$$

Thus
$$v = \frac{l}{\hbar}\frac{dE}{dK}. \tag{4.68}$$

For free electron we have from equation (4.62),

$$E = \frac{\hbar^2 K^2}{2m}$$

or,
$$\frac{dE}{dK} = \frac{\hbar^2 K}{m}.$$

From equation (4.68), we have

$$v = \frac{\hbar K}{m} = \frac{p}{m}. \tag{4.69}$$

i.e., the velocity $v$ is proportional to $K$ and also proportional to momentum $p$ of the particle. In band theory, $E$ is not proportional to $K^2$ as given by equation (4.62). Energy $E$ versus wave vector $K$ according to the band theory is shown in Figure 4.7(a). The curve is a symmetrical curve about $E$ (energy) axis having point of inflection at $P$ and $Q$, the value of $\frac{dE}{dK}$ is maximum but negative at $P$ and positive at $Q$. At the points $A$, $O$ and $B$, the slope $\frac{dE}{dK}$ will be zero. The value of the slope is different for different portion of the curve. The velocity $v$ versus $K$ is obtained from the $E$ versus $K$ diagram which is shown in Figure 4.7(b). The absolute value of the velocity reaches a maximium at $K = K_0$, where $K_0$ corresponds to the inflection point in the $E$-$K$ diagram and beyond this point the velocity decreases with increasing energy which is quite different in behaviour from that of the electrons.

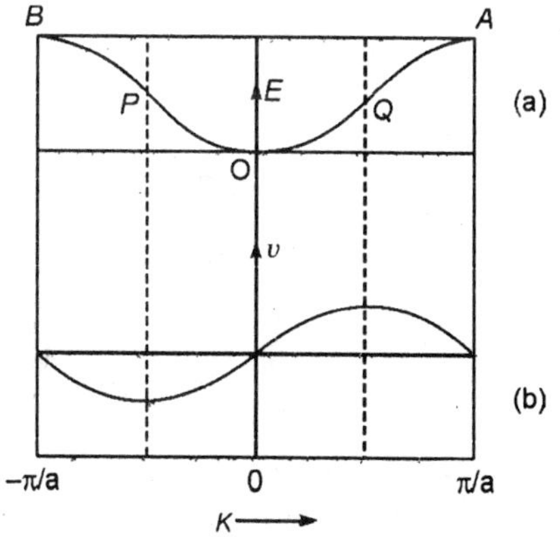


**Figure 4.7** (a) Energy $E$ versus $K$ curve, (b) velocity $v$ versus $K$ curve

**The effective mass of an electron in crystals**

The electrons in crystals are not free completely but they are interacting with the periodic potential of the crystal lattice. So the wave particle motion is different from that of the electron in free space. Now we will investigate the influence of the external electric field $\varepsilon$ on the motion of an electron present in the Brillouin zone under consideration and we assume only one electron which is initially in the state $K$ in the Brillouin zone so that the Pauli's exclusion principle is not violated. Let the field acting on the electron for a time $dt$ so that the electron gains energy $dE$ given by

$$dE = e\varepsilon\, dx = e\varepsilon\, v dt. \tag{4.70}$$

where $v$ is the velocity of electron and can be represented as $v = \frac{d\omega}{dK}$, where $\omega$ is the angular velocity of the wave. But we know,

$$E = h\nu = \hbar\,\omega$$

or,
$$\frac{d\omega}{dk} = \frac{1}{\hbar}\frac{dE}{dk}$$

or,
$$v = \frac{1}{\hbar}\frac{dE}{dK}. \tag{4.71}$$

The equation (4.70) with the help of (4.71) becomes

$$dE = \frac{e\varepsilon}{\hbar}\left(\frac{dE}{dK}\right)dt$$

or,
$$\frac{dE}{dK}dK = \frac{e\varepsilon}{\hbar}\left(\frac{dE}{dK}\right)dt$$

or,
$$\frac{dK}{dt} = \frac{e\varepsilon}{\hbar}. \tag{4.72}$$

The equation (4.72) shows that the rate of change of crystal momentum is equal to the impressed force. The acceleration of the electron is given by equation (4.67) as

$$a = \frac{dv}{dt} = \frac{l}{\hbar}\left(\frac{d^2E}{dK^2}\right)\frac{dK}{dt}. \tag{4.73}$$

From equation (4.72) the equation (4.73) becomes

$$a = \frac{eE}{\hbar^2} = \left[\frac{d^2E}{dK^2}\right]. \tag{4.74}$$

For a free electron of mass $m$ the acceleration is given by

$$a = \frac{eE}{m}. \tag{4.75}$$

Taking $m^*$ as effective mass such that $m^* = \frac{eE}{a}$ as *effective mass of the electron* in the crystal we have

$$m^* = \left(\frac{\hbar^2}{\frac{d^2E}{dK^2}}\right) \tag{4.76}$$

Thus the effective mass of an electron is determined by the value $\left(\frac{d^2E}{dK^2}\right)$ and is a function of $K$ and it is not constant.

It is frequently convenient to introduce a factor $f_K$ such that

$$f_K = \frac{m}{m^*} = \left(\frac{m}{\hbar^2}\right)\left(\frac{d^2E}{dK^2}\right) \tag{4.77}$$

The expression (4.77) is a measure for the extent to which an electron in the $K$-state behaves as a free electron. If $f_K = 1$, the electron behaves as a free electron and if $m^*$ is large and $f_K$ is small, then the electron behaves as a free particle.

Physical significance of effective mass provides a satisfactory description of the charge carriers in a crystal. When the crystals have partially filled energy band, the conduction takes place through electrons but in crystals which are nearly full of holes act as carriers to produce conduction.

## 4.5 THE DISTINCTIONS BETWEEN METALS, INSULATORS, AND INTRINSIC SEMICONDUCTORS

The distinction between metals, insulators and intrinsic semiconductors can be discussed on the basis of band theory of solids. Let us consider an energy band filled with electrons up to a certain value $K_1$ as shown in Figure 4.8. As far as the influence of an external field is concerned we shall be interested in the effective number of free electrons present in the band which gives the conductivity associated with the band. The effective number of free electrons in the band effective for conduction is given by

$$N_{eff} = \sum f_K\,, \tag{4.78}$$

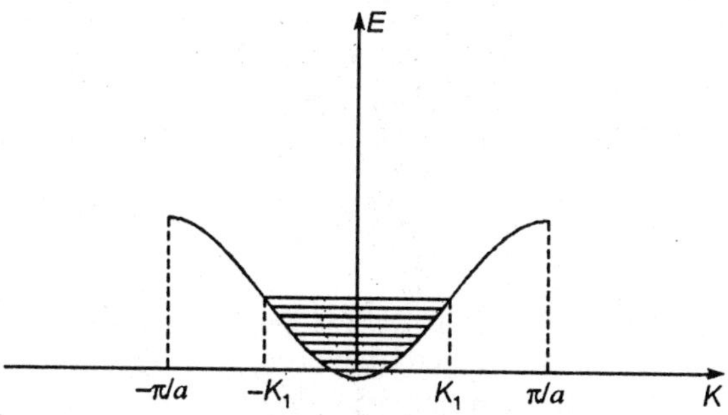


**Figure 4.8** Energy band filled up to states $K_1$ at $T = 0$

where summation extends over occupied states in the band. The number of states in the range $dK$ excluding the spin for one-dimensional lattice of length $L$ is given by the equation (4.65) as

$$dn = \frac{L}{2\pi}dK \tag{4.79}$$

Since two electrons are occupied, each of these states, i.e., within the limits $-K_1$ to $+K_1$,

$$N_{eff} = \frac{1}{\pi}\int_{-K_1}^{K_1} f_K\,dK \tag{4.80}$$

With the help of equation (4.77) the equation (4.80) becomes

$$N_{eff} = \frac{2Lm}{\pi\hbar^2}\int_0^{K_1}\left(\frac{d^2E}{dK^2}\right)dK$$

or, $$N_{eff} = \frac{2Lm}{\pi\hbar^2}\left(\frac{dE}{dK}\right)_{K=K_1} \tag{4.81}$$

We draw the following conclusions from equation (4.81):

1. The effective number of electrons in a completely filled band vanish because $\frac{dE}{dK}$ vanishes at the top of the band.
2. The effective number of electrons reach to maximum value corresponding to the point of inflection because $\frac{dE}{dK}$ is a maximum.

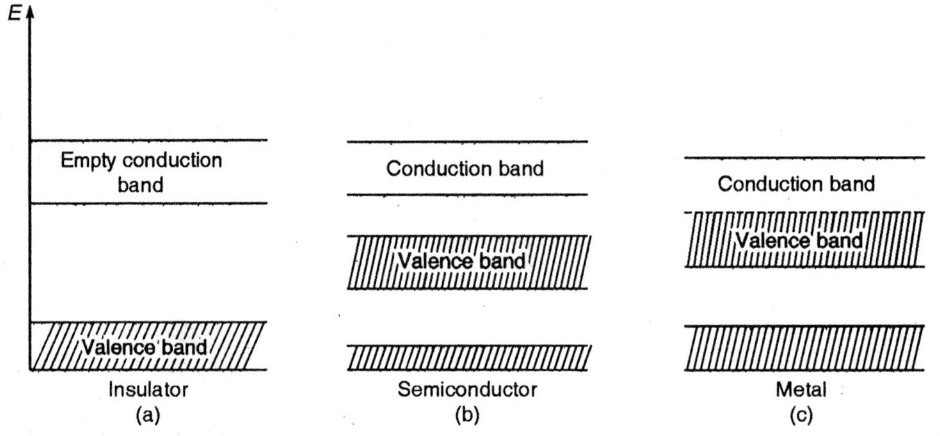


**Figure 4.8** Electron distribution at absolute zero in an insulator, intrinsic semiconductor and metal. Electrons are occupied in the shaded regions

It follows from the above discussion that a solid for which a certain number of energy band is completely filled and the other bands being completely empty is an *insulator* as in Figure 4.8(a). On the other hand, a solid containing an energy band having incompletely filled has *metallic* character, Figure 4.8(c). The Figure 4.8(a) can occur at absolute zero. At temperature greater than absolute zero some electrons from upper filled band will be excited to the next empty band (conduction band) and conduction is possible. If the energy gap is large being the order of several eV, i.e., approximately 7 eV, it will remain as insulator. If the band gap is small, about 1 eV, the number of thermally excited electrons will be appreciable at ordinary temperature and the material is then called *intrinsic semiconductor* [Figure 4.8(b)]. Germanium and silicon are the example of intrinsic semiconductors. The intrinsic semiconductor is insulator at absolute zero and all insulators may be regarded as semiconductors at a temperature greater than absolute zero and in general conductivity of semiconductor increases with the increase of temperature while the conductivity of metal decreases with the rise of temperature.

## 4.6 DRIFT VELOCITY, MOBILITY AND CONDUCTIVITY OF METAL

The conductivities of metals were first explained by Drüde with the presence of mobile electrons. The free electrons in a metal could be treated as an ideal gas of free particle in thermal equilibrium and in non-equilibrium state from application of electric field subject to instantaneous collision. At each collision the electron is assumed to lose any energy it may have acquired from the field and to make a fresh start. These collisions increase the thermal energy of vibration of the positive ion of the metal.

The electrons in metal are in rapid motion because of the thermal energies. An additional velocity is introduced by an external electric field which constitutes the electric current. The net velocity due to the field is called the *drift velocity*. The electron experiences an acceleration

$$a = -\frac{e\varepsilon}{m} \tag{4.82}$$

in the field $\varepsilon$ where $m$ is the mass of the electron and drift velocity $V_d$ is proportional to $\varepsilon$, i.e.,

$$V_d \propto \varepsilon$$

or,

$$V_d = \mu\varepsilon, \tag{4.83}$$

where $\mu$ is a constant called *mobility* and is defined as the drift velocity required by a carrier per unit electric field.

The current density due to electrons is given by

$$J = neV_d, \tag{4.84}$$

where $n$ is the concentration of electrons, $V_d$ is the drift velocity and $e$ is the charge of the electron.

The equation (4.84) can be written with the help of equation (4.83) as

$$J = ne\mu\varepsilon. \tag{4.85}$$

Again, from the Ohm's law we can write,

$$\sigma = \frac{J}{\varepsilon} = ne\mu, \tag{4.86}$$

where $\sigma$ is the *conductivity* of the metals.

## 4.7 HALL EFFECT IN METAL

If a conductivity material is placed in a magnetic field and a current is passed along the length of the conductor, a voltage is developed at right angle to both the current and magnetic field. This phenomenon is known as *Hall effect* and the voltage generated is knwon as *Hall voltage* and its value depends on the magnetic field strength and also on the current passed.

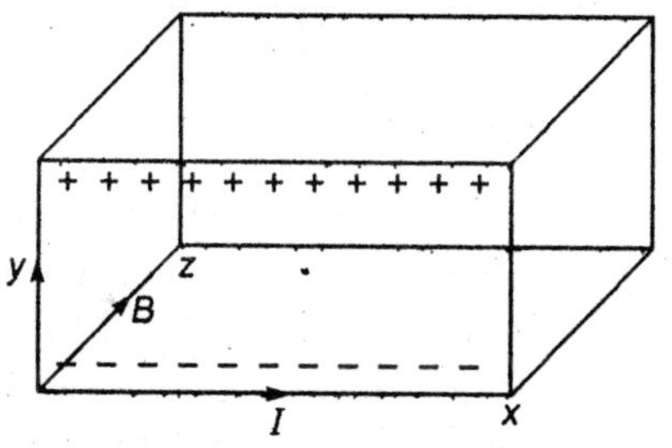


**Figure 4.9** Hall effect in a metal produced by an electric current $I$ along $x$-axis and a magnetic induction $B$ along $z$-axis and force along $y$-axis

Consider a rectangular state of materical carrying an electric current $I$ along $x$-axis and a magnetic induction $B$ along $z$-axis, current carrier will experience a Lorentz force in the negative $y$-direction. When the current is carried by electrons of charge($-e$), then under the influence of magnetic field the electrons will be subjected to a Lorentz force such that the lower surface collects a negative charge and upper surface a positive charge. A stationary state is obtained finally in which current along $y$-axis vanishes and a field $E_y$ is set up. If the charge carriers were positive the upper surface would be negative and the lower surface positive, i.e., $E_y$ would be reversed. Thus the measurement of the 'Hall voltage' gives the information about the sign of the charge carriers.

The total force exerted on a conduction electron by the combined electric field $E$ and the magnetic induction $B$ is given by Lorentz formula,

$$\vec{F} = -e\left[\vec{E} + \frac{1}{c}\vec{v} \times \vec{B}\right].$$

But in the present case,

$$F_y = -e\left[E_y + \frac{1}{c}v_x B_x\right]. \tag{4.87}$$

In steady state, the force $F_y = 0$, then the equation (4.87) becomes

$$E_y = -\frac{v_x B_x}{c} \tag{4.88}$$

This $E_y$ is called *Hall voltage* and $v_x$ is the *average drift velocity* of the electrons. The current density may be expressed in terms of the number of electrons $n$ per unit volume,

$$I_x = -nev_x$$

or,
$$v_x = -\frac{I_x}{ne}. \tag{4.89}$$

Substituting this value of $v_x$ in equation (4.88), we get

$$E_y = -\frac{I_x B_x}{nec} \tag{4.90}$$

or,
$$\frac{E_y}{I_x B_z} = R_H = -\frac{1}{nec},$$

where $R_H$ is known as *Hall coefficient*.

The coefficient is determined essentially by the sign and density of charge carriers. Hall coefficient is the same sign of the carrier. Thus Hall coefficient is negative if conduction is performed by electrons and positive for positive holes.

## SOLVED PROBLEMS

**1.** Taking the origin at the bottom of the conduction band calculate the crystal momentum for free electron of energy 0.015 eV. Given the effective mass of electron = $0.25m_0$.

**Soln.** Given
$$E = 0.015 \text{ eV} = 0.015 \times 1.6 \times 10^{-19} \text{ J}$$
$$m^* = 0.25m_0 = 0.25 \times 9.1 \times 10^{-31} \text{ kg}$$

We know, $$E = \frac{\hbar^2 K^2}{2m^*}.$$

$\therefore$ crystal momentum $\hbar K = \left(2m^* E\right)^{1/2}$

or, $\hbar K = (2 \times 0.25 \times 9.1 \times 10^{-31} \times 0.015 \times 1.6 \times 10^{-19})^{1/2}$

$= \mathbf{3.30 \times 10^{-26}\ kg\ m/sec.}$

**2.** Using the Kronig-Penney model, show that for $P \ll 1$ the energy of the lowest energy band is

$$E = \frac{\hbar^2 P}{ma^2}.$$

**Soln.** From Kronig-Penney model we know the relation from equation (4.60),

$$P \frac{\sin \alpha a}{\alpha a} + \cos \alpha a = \cos ka$$

For $k = \pm \frac{\pi}{a}$, we get the energy of the lowest band

$$P.\frac{\sin \alpha a}{\alpha a} + \cos \alpha a = \pm 1$$

or, $$P.\frac{\sin \alpha a}{\alpha a} = -\cos \alpha a + 1$$

[Positive sign is considered on the right-hand side.]

or, $$\frac{P}{\alpha a} 2 \sin \frac{\alpha a}{2} \cos \frac{\alpha a}{2} = 2 \sin^2 \frac{\alpha a}{2}$$

For $P \ll 1$ we can write as

$$\frac{P}{\alpha a} = \tan \frac{\alpha a}{2}$$

or, $$\frac{P}{\alpha a} = \frac{\alpha a}{2}$$

or, $$\alpha^2 = \frac{2P}{a^2}$$

or, $$\frac{2mE}{\hbar^2} = \frac{2P}{a^2}$$

$\therefore$ $$E = \frac{\hbar^2 P}{ma^2}$$ **[Proved].**

**3.** The energy wave vector dispersion relation for a one-dimensional crystal of lattice constant $a$ is given by $E(K) = E_0 - \alpha - 3\beta \cos Ka$, where $E_0$, $\alpha$ and $\beta$ are constants, find the value of $K$ at which the velocity of an electron is a maximum.

**Soln.** From equation (4.68), we get the velocity of an electron

$$v = \frac{1}{\hbar}\frac{dE}{dK} = \frac{3\beta a}{\hbar} \sin Ka$$

$v$ is maximum when $\sin Ka = 1$

$$\text{or,} \quad Ka = \frac{\pi}{2}$$

$$\therefore \quad K = \frac{\pi}{2a}.$$

## REVIEW QUESTIONS

1. State and prove the Bloch theorem. **[H.P.U. 1996]**
2. Discuss Kronig-Penney model. Using the model show that the energy spectrum of electron consists of a number of allowed energy band separated by forbidden region. **[Luck. U. 1993]**
3. Using the Kronig-Penney model show that the energy spectrum of electron consists of a number of allowed energy band separated by forbidden regions. Discuss the nature of the energy spectrum in the limiting cases of very weak and very strong potentials. **[Burd. U. 2004, C.U. 2002]**
4. Distinguish between metals, insulators and semiconductors. **[C.U. 1977, 1983]**
5. Define the term "effective mass of an electron". Is it different from the electron mass? What is "crystal momentum"? **[C.U. 1977]**
6. (a) Compare the salient features of the band theory and free electron theory. Discuss the merits and demerits of these theories.
   (b) Discuss the formation of allowed and forbidden energy bands on the basis of the Kronig-Penney model? **[Burd. U. 2002]**
7. Discuss the formation of allowed and forbidden energy bands on the basis of Kronig-Penney model. Set up the appropriate Schrödinger equation and boundary conditions. Use Bloch function as the wave function. **[C.U. 2006]**
8. Discuss the basic ideas of the Kronig-Penny Model of a crystalline solid (You need not solve any equation). What is the key factor in the model which leads to forbidden gaps between bands. **[C.U. 2002]**
9. (a) Discuss qualitatively the basic principles underlying the distinctions between metals, insulators and semiconductors.
   (b) Give the theory of Hall effect in a semiconductor. **[C.U. 1996]**

## PROBLEMS

1. The energy wave vector dispersion relation for a one-dimensional crystal of lattice constant $a$ is given by

   $E(K) = E_0 - \alpha - 2\beta \cos Ka$, where $E_0$, $\alpha$ and $\beta$ are constants.

   (i) Find the value of $K$ at which the velocity of electron is a maximum.
   (ii) Find the difference between the top and bottom of the energy band.

(iii) Obtain the effective mass $m^*$ of the electron at the bottom and at the top of the band.

**[C.U. 2003]**

$$\left[\textbf{Ans. } \text{(i) } K = \frac{p}{2a} \quad \text{(ii) } 4b \quad \text{(iii) } m^* = \frac{\hbar^2}{2\beta a^2} \text{ and } m^* = -\frac{\hbar^2}{2\beta a^2}\right]$$

**2.** The energy near the valence band edge of a crystal is given by

$$E = AK^2,$$

where $A = 10^{-39}$ Jm$^3$. An electron with wave vector $K = 10^{10}\ K_\lambda$ m$^{-1}$ is removed from an orbital in the completely filled valence band. Determine the effective mass, velocity, momentum and energy of the hole.

$$\left[\begin{array}{l}\textbf{Ans. } m_h^* = 5.5 \times 10^{-30}\ \text{kg}, \vec{P}_h = -1.05 \times 10^{-24}\ K_\lambda\ \text{Jsm}^{-1} \\ V_h = -1.9 \times 10^5\ K_\lambda\ \text{ms}^{-1}, E_h = 10^{-19}\ \text{J}\end{array}\right]$$

# 5

# *Magnetic Properties of Material*

## 5.0 INTRODUCTION

Besides mechanical, electrical and optical properties, solid also possesses magnetic properties. It was observed in early 800 B.C. in loadstone which was used for navigation properties. *Metal*, *insulator* and *semiconductor* exhibit magnetism of different nature. It is convenient to group the magnetic properties of solids in the following categories (i) Diamagnetism (ii) Paramagnetism (iii) Ferromagnetism.

The following magnetic terminology is necessary for further study of magnetism in solids. The magnetisation $\vec{M}$ is defined as the magnetic moment per unit volume developed inside a solid. If $\vec{M}$ is the magnetisation induced by the magnetic field $\vec{H}$, the magnetic susceptibility per unit volume is defined as

$$\chi = \frac{\vec{M}}{\vec{H}}. \tag{5.1}$$

Susceptibility $\chi$ is a scalar quantity in isotropic material and it is a tensor in anisotropic substances. In the S.I. system both $\vec{M}$ and $\vec{H}$ are measured in ampere per metre and $\chi$ is a dimensionless quantity. If $\vec{M}$ is expressed in gram-molecule of a substance, then susceptibility is called *molar susceptibility* and is designated as $\chi_m$.

Linear isotropic material can be classified in two groups (i) Diamagnetic (ii) Paramagnetic. In the diamagnetic material $\chi < 0$, while in the paramagnetic material $\chi > 0$.

The magnetic induction or magnetic flux density $\vec{B}$ produced inside a medium by the application of magnetic field $\vec{H}$ is given by the relation:

$$\vec{B} = \mu_0 (\vec{H} + \vec{M}), \tag{5.2}$$

where $\mu_0$ is the magnetic permeability of free space and $\mu_0 = 4\pi \times 10^{-7}$ henry/m.

or, $$\vec{B} = \mu_0 \vec{H}(1 + x) = \mu_0 \mu_r \vec{H} = \mu \vec{H}, \tag{5.3}$$

where $\mu_r = 1 + \chi$ and $\mu = \mu_0 \mu_r = \mu_0 (1 + \chi)$.

$\mu$ is called the absolute permeability of the material. For free space, $\vec{M} = 0, \chi = 0, \mu = \mu_0$ and $\mu_r = 1$, then from equation (5.3), we obtain

$$\vec{B} = \mu_0 \vec{H} \tag{5.4}$$

The unit of magnetic flux is weber (Wb) and the unit of magnetic induction is tesla (T). In S.I. system

$$1 \text{ tesla} = 1 \text{ Wb/m}^2.$$

## 5.1 TYPES OF MAGNETISM

According to the magnetic behaviour, substances are divided into the following categories:

1. Diamagnetism
2. Paramagnetism
3. Ferromagnetism
4. Antiferromagnetism
5. Ferrimagnetism

Diamagnetic substances are feebly repelled by magnet and are observed in solids which do not contain any permanent magnetic moments. In fact, diamagnetism is an atomic property and the existence of a small non-zero magnetic moment in these materials is attributed to the orbital motion of the electrons and the magnetic moment is always directed opposite to the magnetic field.

Paramagnetic substances are feebly attracted by magnets and the magnetic moment is aligned with the direction of the field. Paramagnetic substance does not show hysterisis. In the paramagnetic substance some electronic shell is incomplete when there exists a permanent magnetic moment in these substances.

Ferromagnetic substances are strongly attracted to magnets and have magnetisation in weak field. Ferromagnetic substance can exist in a permanently magnetised state even if there is no external field. The magnetic moments align themselves with the same direction. Ferromagnetic substances lose their magnetisation above a certain critical temperature and then behave as paramagnetic substances.

In antiferromagnetism the crystal may be regarded as composed of two substances with opposite direction of magnetisation which completely neutralise each other. In this type of material susceptibility shows a maximum as a function of temperature.

Ferrimagnetism is similar to antiferromagnetism but oppositely aligned with magnetisations of the two. Sub-lattices are unequal in magnitude, and complete cancellation of magnetic moments does not take place.

## 5.2 LANGEVIN'S THEORY OF DIAMAGNETISM (CLASSICAL THEORY)

The atom contains electrons revolving in complicated orbits and revolving electron in the orbit constitutes an electric current. Since a current flowing in a closed circuit behaves like a magnetic shell having magnetic moment $\mu = SI$, where $S$ is the area of the current loop and $I$ is the current strength. If the orbit is assumed to be circular of radius $r$, then $S = \pi r^2$, so $\mu = \pi r^2 I$. If the electron revolves in the circular orbit with angular velocity $\omega_0$, the current in the loop $-\frac{e\omega_0}{2\pi}$, hence the magnetic dipole moment associated with electron orbit is $\mu = -\frac{e\omega_0}{2\pi}\pi r^2$.

or, $$\mu = -\frac{e\omega_0 r^2}{2}.$$

The magnetic moment has a direction at right angle to the plane of the paper and it is opposite to the direction of the orbital angular momentum $\vec{p}_L$ of the electron.

The magnitude $\vec{p}_L$ is given by

$$\vec{p}_L = mr^2\omega_0, \tag{5.5}$$

where $m$ is the mass of the electron.

The ratio of magnetic moment $m$ to the orbital angular momentum is

$$\frac{\mu}{p_L} = -\frac{e\omega_0 r^2}{2} \cdot \frac{1}{mr^2\omega_0} = -\frac{e}{2m}. \tag{5.6}$$

This ratio is known as *gyromagnetic ratio.*

$$\therefore \qquad \vec{\mu} = -\frac{e}{2m}\vec{p}_L. \tag{5.7}$$

The formula in equation (5.7) is known as *Larmor's formula*.

Now consider the influence of the magnetic field on the motion of an electron in an atom quantitatively. The magnetic field produces torque $\vec{\mu} \times \vec{H}$ on the dipole; so according to classical mechanics we may write,

$$\frac{d}{dt}(\vec{p}_L) = \vec{\mu} \times \vec{H} = -\frac{e}{2m}(\vec{p}_L \times \vec{H}). \tag{5.8}$$

This is the equation of motion of a vector $\vec{p}_L$ precessing about $\vec{H}$ with an angular frequency

$$\omega_L = \frac{eH}{2m} \tag{5.9}$$

where $\omega_L$ is called *Larmor frequency*. This is seen in Figure 5.1. The Larmor precessional frequency

$$\nu_L = \frac{\omega_L}{2\pi} = \frac{eH}{4\pi m}. \tag{5.10}$$

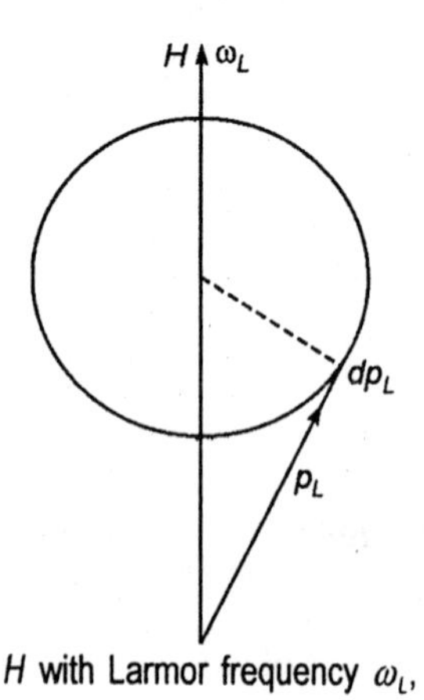


**Figure 5.1** Precession of $p_L$ and $H$ with Larmor frequency $\omega_L$, $dp_L$ is perpendicular to the plane of $\vec{p}_L$ and $\vec{H}$

From equation (5.8), we have

$$dp_L = \vec{\omega}_L \times \vec{p}_L\, dt = \frac{e}{2m} \vec{H} \times \vec{p}_L\, dt\,. \tag{5.11}$$

This derivation is based on the assumptions that $\vec{p}_L$ is independent of $\vec{H}$, i.e., it is assumed that the orbit is not deformed under the influence of the magnetic field. As a result of charge of electron the precession produces an induced magnetic moment with a component opposite to that of $\vec{H}$.

$$\therefore \qquad (\mu_{ind})_H = -\frac{e}{2m}\Delta p_L = -\frac{e}{2m} m\omega_L < \rho^2 > = -\frac{e\omega_L}{2} < \rho^2 >$$

$$= -\frac{e\omega_L}{2} < \rho^2 >, \tag{5.12}$$

where $< \rho^2 >$ is the mean square radius of projection of the orbit on a plane perpendicular to $\vec{H}$. The quantity $m\omega_L < \rho^2 >$ is the change in angular momentum of the electron due to the precessional motion induced by the magnetic field. When the treatment is extended to a solid containing $N$ atoms in unit volume containing $Z$ electrons, the diamagnetic susceptibility

$$\chi_{dia} = -NZ \frac{e^2}{4m} < \rho^2 >$$

or,

$$\chi_{dia} = -NZ \frac{e^2}{6m} < r^2 >, \tag{5.13}$$

where it is assumed that the charge distribution of the atom is spherically symmetrical so that

$$< r^2 > = \frac{3}{2} < \rho^2 >.$$

[For spherically symmetric charge distribution

$$< x^2 > = < y^2 > = < z^2 >$$

$$< r^2 > = < x^2 > + < y^2 > + < z^2 >$$

or,

$$< x^2 > = < y^2 > = < z^2 > = \frac{1}{3} < r^2 >$$

and

$$< r^2 > = < x^2 > + < y^2 > = \frac{2}{3} < r^2 >]$$

$\chi_{dia}$ is negative. Since the mean distance of the orbital electron is of the order of $10^{-10}$ m,

$$< r^2 > \sim 10^{-20} \text{ m, } N = 10^{28} \text{ per m}^3$$

$$\therefore \qquad \chi_{dia} = -10^{-7} Z \tag{5.14}$$

This is the classical Langevin's theory of diamagnetism and shows that diamagnetic susceptibility is independent of temperature.

## 5.3 LANGEVIN'S THEORY OF PARAMAGNETISM (CLASSICAL THEORY)

Paramagnetism occurs in the substances which have permanent magnetic moments. In the absence of magnetic field these magnetic moments are oriented randomly and no net magnetisation is produced. Langevin was the first to give an electron theory of paramagnetism.

Let $\mu$ be the magnetic dipole moment of each atom. The mutual interaction between the magnetic dipoles is negligible in the Langevin's theory. Hence, in the presence of magnetic induction $\vec{B}$, these dipoles tend to orient themselves in the direction of the field, and magnetic potential energy of each dipole is

$$E = -\vec{\mu}.\vec{B} = -\mu B \cos\theta, \tag{5.15}$$

where $\theta$ is the angle between $\vec{\mu}$ and $\vec{B}$. The atomic dipoles are aligned with the direction of the field in order to minimise their energy. The random thermal motion of the molecules oppose this tendency. Under the influence of these two opposing tendencies each atomic magnetic dipole orients itself in the direction of magnetic induction and reaches an equilibrium position. The projection of the dipole moment in the direction of $\vec{B}$ will be $\mu$ cos $\theta$. The magnetic moment induced in the substance by the external magnetic induction will be equal to the algebraic sum of the projection of the moments of all the dipoles present in a unit volume.

Using Maxwell-Boltzmann distribution law, the probability of dipole having orientation between $\theta$ and $\theta + d\theta$ is given by

$$W(\theta) \propto \exp\left(-\frac{\varepsilon}{KT}\right) d\Omega\, d\theta, \tag{5.16}$$

where $d\Omega$ is the solid angle and $d\Omega = 2\pi \sin\theta\, d\theta$ and $K$ is the *Boltzmann constant* and $T$ is the absolute temperature.

Using equation (5.15), (5.16) may be written as

$$W(\theta)\, d\theta = C \exp\left(-\frac{\mu B \cos\theta}{KT}\right) 2\pi \sin\theta\, d\theta, \tag{5.17}$$

where $C$ is a constant.

Hence the average component of the magnetic moment of each atom in the field direction is

$$M = n < \mu \cos\theta >, \tag{5.18}$$

where $n$ is the number of atoms per unit volume. $<\mu \cos\theta>$ is the mean value of projection of the magnetic moment in the field direction.

$$<\mu\cos\theta> = \frac{\int_0^\pi \mu \cos\theta\, W(\theta)\, d\theta}{\int_0^\pi W(\theta)\, d\theta}$$

$$= \frac{\mu \int_0^\pi \exp\left(\frac{\mu B_0 \cos\theta}{KT}\right) \cos\theta \sin\theta\, d\theta}{\int_0^\pi \exp\left(\frac{\mu B_0 \cos\theta}{KT}\right) \sin\theta\, d\theta}. \tag{5.19}$$

Let $$x = \cos\theta \quad \text{and} \quad y = \frac{\mu B_0}{KT}.$$

$\therefore$ $$dx = -\sin\theta\, d\theta$$

$$<\mu\cos\theta> = \frac{\mu\int_{-1}^{1} x\exp(xy)\,dx}{\int_{-1}^{+1}\exp(xy)\,dx}$$

$$= \mu\left\{\frac{\exp(y)+\exp(-y)}{\exp(y)-\exp(-y)} - \frac{1}{y}\right\}$$

$$= \mu\left\{\coth(y) - \frac{1}{y}\right\}. \tag{5.20}$$

From equations (5.18) and (5.20), we can write,

$$M = n\mu\, L(y), \tag{5.21}$$

where $L(y) = \left\{coth(y) - \frac{1}{y}\right\}$ is called *Langevin function*. The variation of $L(y)$ with $y$ is shown in Figure 5.2. For small $y$, i.e., $y << 1$, we can expand coth $(y) = \frac{1}{y} + \frac{y}{3}$.

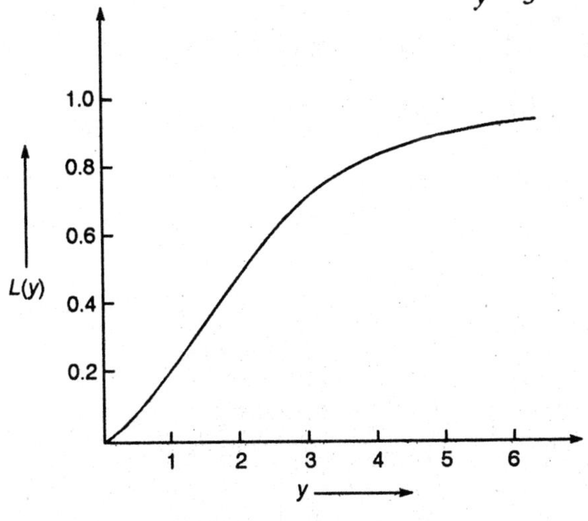


**Figure 5.2** Variation of $L(y)$ with $y$

$\therefore$ $$L(y) = \frac{y}{3} + \frac{\mu B_0}{3\,KT} \tag{5.22}$$

and $$M = \frac{n\mu^2 B_0}{3\,KT}. \tag{5.23}$$

The paramagnetic susceptibility,

$$\chi_{para} = \frac{\mu_0 M}{3KT} = \frac{n\mu_0\mu^2}{3KT}. \tag{5.24}$$

The equation (5.24) shows that the susceptibility is inversely proportional to temperature. The equation (5.24) can be written as

$$\chi_{para} = \frac{C}{T}, \tag{5.25}$$

where $C = \frac{n\mu_0\mu^2}{3K}$ is a constant.

The expression (5.25) is called the *Curie law* and $C$ is called *Curie constant*.

For large value of $B_0$ and low temperature, the $y$ becomes large. For $y >> 1$, $L(y)$ approaches unity and then equation (5.21) becomes

$$M = n\mu. \tag{5.26}$$

Thus at a very low temperature all the atomic dipoles are aligned with the direction of the field and $M$ is called the *saturation magnetisation*.

## 5.4 QUANTUM THEORY OF PARAMAGNETISM

The Langevin's classical theory of paramagnetism provides a good description of the temperature dependence of the susceptibility of an ideal paramagnetic gas but cannot explain the origin of the atomic moment. In Langevin's theory it is assumed that the permanent magnetic moment of a free atom or ion can possess any orientation with respect to magnetic field. But according to quantum theory these magnetic moments are quantized, the magnetic dipole moment $\vec{\mu}$ and its component $\mu_B$ in the direction of the applied field cannot have arbitrary values. Confining attention to a free atom or ion we associate a quantum number $\vec{J}$, where $|\vec{J}| = \sqrt{J(J+1)}$ with the total resolved component of the angular moment as

$$\vec{\mu} = -g\mu_B\vec{J} \tag{5.27}$$

The quantity $\mu_B$ is called *Bohr magneton* and is equal to $\frac{e\hbar}{2m}$ in S.I. system and $\vec{J}\hbar = \vec{L}\hbar + \vec{S}\hbar$, where $\vec{L}$ is the orbital and $\vec{S}$ is the spin quantum number and $g$ is the *Lande's g factor* and is given by

$$g = 1 + \frac{J(J+1) + S(S+1) - L(L+1)}{2J(J+1)}. \tag{5.28}$$

The special quantization of the magnetic moment can assume discrete set of values in the field direction as

$$-g\mu_B J,\ -g\mu_B\,(J-1),\ -g\mu_B\,(J-2),\ \ldots,\ 0,\ \ldots,\ g\mu_B\,(J-2),\ g\mu_B\,(J-1),\ g\mu_B J. \tag{5.29}$$

In all there are $(2J+1)$ possible orientations. We describe the resolved value of $J$ by magnetic quantum number $m$ so that arbitrary resolved component of magnetic moment along the field direction is given by $g\mu_B m$.

An atom with magnetic quantum number $m$ acquires a potential energy in the presence of a magnetic field $B_0$.

$$\varepsilon_m = -g\mu_B m B_0 \tag{5.30}$$

Consider a unit volume of a paramagnetic material containing a total of $N$ atoms; the statistical average of the magnetic moment component per atom along $B_0$ is

$$M = N\frac{\sum_{-J}^{+J} mg\mu_B \exp(mg\mu_B B_0 / KT)}{\sum_{-J}^{+J} \exp(mg\mu_B B_0 / KT)}. \tag{5.31}$$

Writing $\beta = (g\mu_B B_0/KT)$ the equation (5.31) becomes

$$M = N\frac{\sum_{-J}^{+J} mg\mu_B \exp(m\beta)}{\sum_{-J}^{+J} \exp(m\beta)}. \tag{5.32}$$

Now we consider two cases:

(i) We assume that $\beta$ is small and magnetic field is not too high, i.e., $B_0 < 1$ T and at higher temperature we can write,

$$\exp(\beta m) = 1 + \beta m. \tag{5.33}$$

So we can write the equation (5.32) as

$$M = N\frac{\sum_{-J}^{+J} g\mu_B m(1+\beta m)}{\sum_{-J}^{+J}(1+\beta m)}. \tag{5.34}$$

Now we need the following sums

$$\sum_{-J}^{+J} m = 0 \tag{5.35}$$

$$\sum_{-J}^{+J} 1 = \sum_{-J}^{+J} m^0 = 2J+1 \tag{5.36}$$

and

$$\sum_{-J}^{+J} m^2 = \frac{1}{3}J(J+1)(2J+1). \tag{5.37}$$

With the help of equations (5.35), (5.36) and (5.37) the equation (5.34) becomes

$$M = Ng^2\mu_B{}^2 B_0 \frac{J(J+1)}{3KT} \tag{5.38}$$

or,
$$\frac{M}{B_0} = Ng^2\mu_B{}^2 \frac{J(J+1)}{3KT} \tag{5.39}$$

or,
$$\chi_{para} = \frac{\mu_0 M}{B_0} = \frac{\mu_0 N\mu_B{}^2}{3KT} g^2 J(J+1) \tag{5.40}$$

or,
$$\chi_{para} = \frac{\mu_0 N P_{\text{eff}}^2 \mu_B{}^2}{3KT} = \frac{C}{T}, \tag{5.41}$$

where $P_{\text{eff}}^2 = g^2 J(J+1)$ is the *effectvie number of Bohr magneton*. $P_{\text{eff}}$ can be determined by measuring $\chi_{\text{para}}$. The expression is identical with the classical expression (5.24) substituting

$$\mu^2 = P_{\text{eff}}^2\ \mu_B^2. \tag{5.42}$$

Implicit in the above derivation we assume that all the atoms are in the same spectroscopic state, i.e., all have the same $J$ value.

(ii) At low temperature, say about 4 °K, and strong magnetic fields the conditions imposed (i) may not be satisfied since $\beta$ is a small quantity and $m\beta$ is not smaller than unity and then it is not possible to make series expression of the exponential terms in the expression (5.32). After some algebraic manipulation the expression (5.32) becomes

$$M = NgJ\mu_\beta B_J(x), \tag{5.43}$$

where $x = \dfrac{g\,J\mu_B\,B_0}{KT}$ and $B_J(x)$ is the Brillouin function defined by

$$B_J(x) = \frac{2J+1}{2J}\coth\left[\frac{(2J+1)x}{2J}\right] - \frac{1}{2J}\coth\left(\frac{x}{2J}\right). \tag{5.44}$$

For $x << 1$, we have

$$\coth x = \frac{1}{x} + \frac{x}{3} \tag{5.45}$$

$$\therefore \quad B_J(x) = \frac{(2J+1)}{2J}\left[\frac{2J}{(2J+1)\,x} + \frac{(2J+1)}{3.2J}x\right] - \frac{1}{3.2J}\left[\frac{2J}{x} + \frac{x}{3.2J}\right]$$

$$= \frac{(J+1)\,x}{3J} \tag{5.46}$$

Thus the susceptibility becomes

$$\chi_{para} = \frac{\mu_0 M}{B_0} = \frac{\mu_0 N\, J(J+1) g^2 \mu g^2}{3KT} \tag{5.47}$$

$$= \frac{\mu_0 N\, P_{eff}^2\, \mu_B^2}{3KT}. \tag{5.48}$$

For $x >> 1$, i.e., for high field and low temperature, coth $x \cong 1$.

$$\therefore \qquad B_J(x) \cong 1. \tag{5.49}$$

The equation (5.43) becomes

$$M = Ng\, J\mu_B. \tag{5.50}$$

This result indicates the state of the magnetic saturation, i.e., all the dipoles align with the magnetic field $B_0$. For J→ ∞, i.e., for a large number of allowed orientation of magnetic dipole we have

$$\coth \frac{x}{2J} \to \frac{2J}{x}$$

and

$$\coth\left(\frac{2J+1}{2J}\right)x = \coth\left(1+\frac{1}{2J}\right)x \to \coth x.$$

$$\therefore \qquad B_J(x) \to \coth x - \frac{1}{x} = L(x). \tag{5.51}$$

Thus quantum results [equation (5.51)] approach the classical ones [equation (5.21)] and equivalent to *Langevin's function*.

The equation (5.48) can be used to calculate the paramagnetic susceptibility of a solid.

The permanent magnetic dipole moments of the atoms arise from the orbital and spin motion of the electrons. Only those atoms which have partially filled electronic sub-shells have permanent magnetic moment. So there is no contribution to the magnetic moment of the atoms from completely filled shell.

Paramagnetic susceptibility of a solid at room temperature is very low, about $10^{-7}$. It increases 100 times at low temperature at 1 °K.

*J* value in equation (5.41) may be determined in high temperature approximately by applying *Hund's rule*. Hund used the vector atom model which postulates that it is always subject to Pauli's exclusion principle.

(i) The electron spin *S* is maximised.

(ii) The orbital momentum *L* is maximised consistently with (i).

(iii) $J = \begin{cases} |L-s| & \text{when a shell is less than half-filled} \\ |L+s| & \text{when a shell is more than half-filled.} \end{cases}$

When a shell is exactly half-filled, $L = 0$. Therefore, $J = S$, since $\Sigma m = 0$.

The order of magnitude of the paramagnetic susceptibility of a solid per m³ may be estimated from equation (5.41). Since $\mu_B \simeq 10^{-23}$ J/T, $n$ ~ $10^{28}/\text{m}^3$, we obtain $\chi \approx \frac{1}{300T}$. At room temperature $T = 300$ °K, $\chi \simeq 10^{-5}$ and at 1°K, $\chi \simeq 10^{-2}$.

Apart from the electronic paramagnetism the atoms also have nuclear paramagnetism but *nuclear magnetic moments* have magnitude less than one-thousandth of the atomic magnetic moments. Hence, the electronic paramagnetism is more important than *nuclear paramagnetism*.

## 5.5 FERROMAGNETISM

### (a) Introduction

Ferromagnetic substances are those which can be magnetised to saturation by application of small field and retain magnetism even when the external field has been removed. Ferromagnetic substances are strongly attracted by magnets. They are crystalline in nature and are a few in number, viz., Fe, Co, Ni, Cd, Dy, etc., and their alloys. Recently more ferromagnetic compounds have been discovered, none of whose components are ferromagnetic. Compounds are MnAs, AgFe, $CrO_2$, etc. A ferromagnetic solid exhibits a magnetic moment in the absence of a magnetic field and is called *spontaneous magnetization.* This spontaneous magnetization exists below a certain critical temperature known as *ferromagnetic Curie temperature* $\theta_C$. A spontaneous moment suggests that electron spins and magnetic moments are arranged in regular manner and all spin arrangements sketched in Figure 5.3 have a spontaneous magnetic moment usually called *saturation moment*.

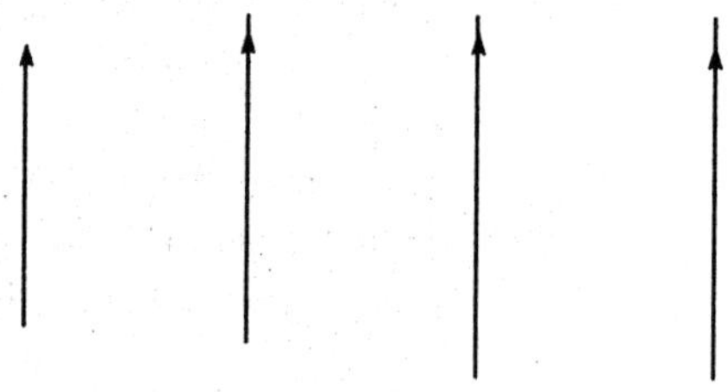

**Figure 5.3** Pattern of spin arrangement in simple ferromagnetic material

### (b) Weiss theory of Ferromagnetism

The theory of Ferromagnetism was first developed by P. Weiss in 1907. It is known as *molecular field theory*. The theory is based on two main assumptions:

(i) Ferromagnetic substances are exhibited by crystalline solids. It is believed that in ferromagnetic substances there are small regions smaller than grain sizes containing $10^{17}$ to $10^{21}$ atoms in which all elementary axes are aligned in the same direction. Such fully magnetized region forms 'domain'. These atomic magnets owe their origin to the spin of electrons in the unfilled inner shell of the atom. The formation of the domain depends on the strong interatomic force in the ferromagnetic crystal. A ferromagnetic mateial of small dimensions consists of a number of domain which is spontaneously magnetised. The magnitude of the *spontaneous magnetization* of the ferromagnetic material is the result of the vector sum of the magnetic moments of these domains.

(ii) The spontaneous magnetization of the microscopic domain arises due to the existence of molecular fields within the material which tends to produce alignment with the atomic dipoles.

Weiss modified the Langevin's theory by postulating that the effective magnetic field $B_{eff}$ on an atom or ion becomes

$$B_{eff} = B + B_E, \tag{5.52}$$

where $B$ is the applied field and $B_E$ is an exchange field and also called *molecular field* or *Weiss field* and it is assumed proportional to the magnetisation $M$ of each domain, i.e.,

$$B_E = \lambda M, \tag{5.53}$$

where $\lambda$ is a constant and is called *Weiss field constant* or *molecular constant* or simply *Weiss constant* and it is independent of temperature.

In order to develop the theory of ferromagnetism we shall use the qunatum theory of paramagnetism rather than the classical Langevin used by Weiss in his original article.

Consider a ferromagnetic solid containing $N$ atoms per unit volume each with a total angular momentum quantum number $J$ which includes the total orbital quantum number $L$ and total spin quantum numbers $S$ such that $J = L + S$. According to the result of quantum theory of paramagnetism, we can write the expression (5.43) as

$$M = Ng\, J\, \mu_B\, B_J(x), \tag{5.54}$$

where $$x = \frac{g\, \mu_B J\, B_{eff}}{KT}$$

and $B_J(x)$ is the *Brillouin function* defind by

$$B_J(x) = \frac{2J+1}{2J} \coth\left[\frac{(2J+1)x}{2J}\right] - \frac{1}{2J} \coth\left(\frac{x}{2J}\right) \tag{5.55}$$

$$x = \frac{g\, \mu_B\, JB_{eff}}{KT} = \frac{gJ\, \mu_B}{KT}\,(B + \lambda M). \tag{5.56}$$

In case of spontaneous magnetization $B = 0$, the equation (5.55) becomes

$$x = \frac{gJ\, \mu_B\, \lambda M}{KT} \tag{5.57}$$

and $$M = \frac{x\, KT}{g\, J \mu_B \lambda}. \tag{5.58}$$

For $T \to 0$, $x \to \infty$, then according to equation (5.49), $B_J(x) \to 1$, the magnetic moment aligns themselves parallel to the field and then magnetization $M$ becomes saturation magnetism $M_S$. Therefore, from equation (5.53) we can write,

$$M_S(0) = Ng\, J\mu_B. \tag{5.59}$$

From equations (5.58) and (5.59) we can write,

$$\frac{M}{M_S(0)} = \frac{x\, KT}{\lambda\, Ng^2 \mu_B^2 J^2}. \tag{5.60}$$

Again, from equations (5.54) and (5.58) we can obtain

$$\frac{M}{M_S(0)} = B_J(x). \tag{5.61}$$

The two equations (5.60) and (5.61) can be used to plot $M/M_S(0)$ versus $x$ as shown in Figure 5.4 for $T > \theta_C$, $T = \theta_C$ and $T < \theta_C$. $M(T)/M_S(0)$ vs. $x$ represent a straight line, the slope of the line being proportional to $T$. From the Figure 5.4, it follows that for $T < \theta_C$ one obtains a nonvanishing value of $M$ even for zero external field, hence it corresponds to spontaneous magnetization. Spontaneous magnetization decreases with increase of temperature and for $T \geq \theta_C$, the spontaneous magnetization vanishes. The temperature $\theta_C$ is known as ferromagnetic Curie temperature. The graph of $M/M_S(0)$ against $T/\theta_C$ is a universal graph applicable to all material which is shown in Figure 5.5.

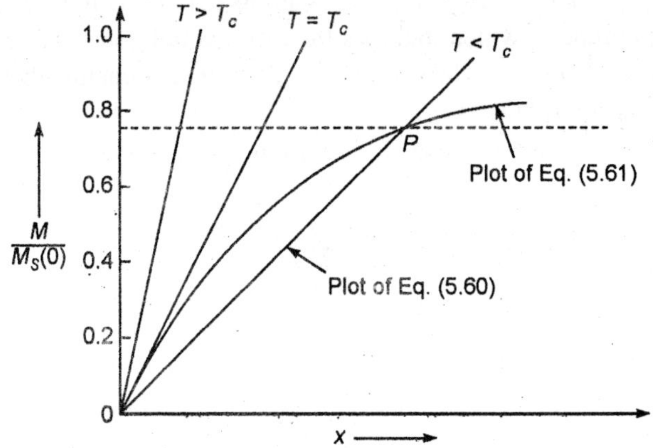


**Figure 5.4** Variation of $M/M_s(0)$ for $x$ with a ferromagnetic material at different temperature

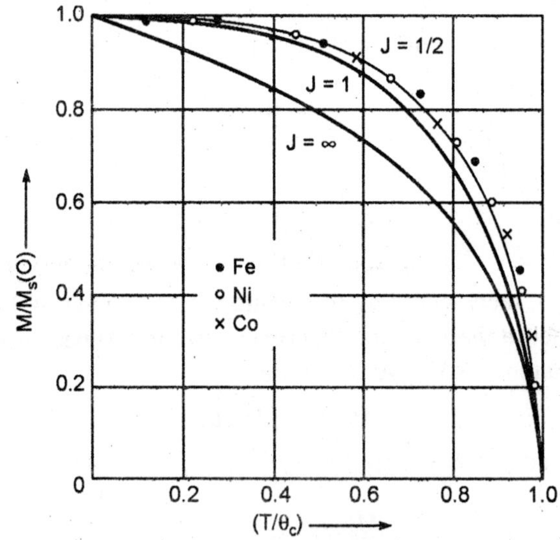


**Figure 5.5** Plot of $M/M_s(0)$ against $(T/\theta_c)$ for Fe, Ni and Co. The curve for $J = 1/2$, $J = 1$ and $J = \infty$ are those obtained from equations (5.60) and (5.61)

Now for $x << 1$ we can write from equation (5.46),

$$B_J(x) = \frac{(J+1)x}{3J}.$$

The equation (5.54) with the help of equation (5.46) becomes

$$M = Ng\ J\ \mu_B\ \frac{(J+1)x}{3J}. \tag{5.62}$$

where $x$ from equation (5.56) can be written as

$$x = \frac{gJ\ \mu_B(B+\lambda M)}{KT}$$

or,
$$M = \frac{Ng^2\ \mu_B^2(B+\lambda M)}{KT}\ \frac{J(J+1)}{3}. \tag{5.63}$$

Putting $C_1 = \dfrac{Ng^2\mu_B^2\ J(J+1)}{3K}$ the equation (5.63) can be written as

$$M_1 = \frac{C_1\ (B+\lambda M)}{T} \tag{5.64}$$

This gives the susceptibility $\chi$ as

$$\chi = \frac{\mu_0 M}{B} = \frac{C}{T-\theta_c}, \tag{5.65}$$

where
$$C = \frac{\mu_0\theta_c}{\lambda} \tag{5.66}$$

and
$$\theta_C = C_1\ \lambda = \frac{\lambda Ng^2\ J(J+1)}{3K}. \tag{5.67}$$

The expression in (5.65) is known as *Curie-Weiss law.* The equation (5.65) describes satisfactorily the temperature dependence of susceptibility in paramagnetic region if the temperature is well above *Curie temperature.*

**(c) Antiferromagnetism**

Antiferromagnetism originates if the spin vectors of the neighbouring atoms are antiparallel arrangement as in Figure 5.6 or when a negative exchange interaction results from antiparallel spin moment in adjacent atoms. An example of antiferromagnetic substance is MnO. The

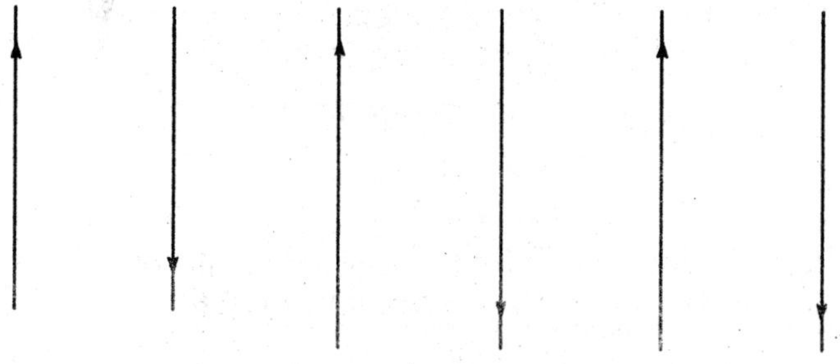

**Figure 5.6** Neighbouring antiparallel spin moments of sub-lattice producing antiferromagnetism

susceptibility versus temperature curve shows a sharp maximum at a temperature which is known as *Neel temperature* $T_N$ as in Figure 5.7.

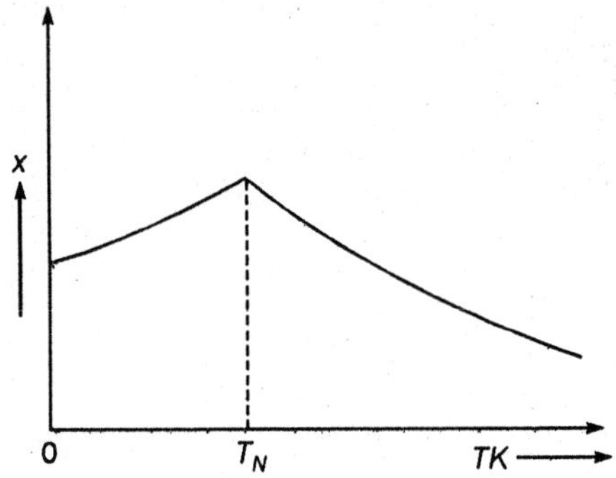


**Figure 5.7** Susceptibility versus temperature of a antiferromagnetic substance

Above the Neel temperature susceptibility decreases continuously and then the substance behaves like an ordinary paramagnetic substance. In the absence of the external magnetic field and below Neel temperature the net magnetic moment is zero in antiferromagnetic substance. The susceptibility $\chi$ of antiferromagnetic substance is given by

$$\chi = \frac{C}{T+\theta},$$

where $\theta$ is called the *asymptotic Curie temperature* and $T > T_N$, $C$ is the Curie constant.

### (d) Ferrimagnetism (ferrites)

If in some substance the oppositely aligned magnetizations of the two sub-lattices have different magnitudes they cannot completely compensate for each other. Such a substance is said to exhibit ferrimagnetism. The most common example is ferrous ferrite $Fe_3O_4$ or FeO $Fe_2O_3$. Arrangement of spin vectors of fe 'tes are shown in Figure 5.8. The most important property of ferrites is that their d.c. resistivity is $10^4$ to $10^{11}$ times as great as that of iron. They are used in ferromagnets for the use of high frequencies where eddy current losses are a serious problem. The high resistivity of ferrites are very useful in microwave device. They are also used for memory storage in computers.

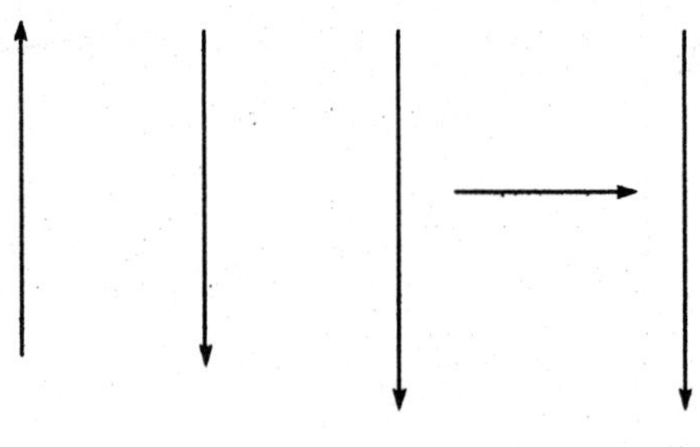

**Figure 5.8** Neighbouring spin moments of sub-lattice producing ferromagnetism

## 5.6 DIFFERENCES AMONG FERROMAGNETIC, PARAMAGNETIC AND DIAMAGNETIC SUBSTANCES

**The differences between the three classes of magnetic substance are given in a tabular form:**

| Ferromagnetic substance | Paramagnetic substance | Diamagnetic substance |
|---|---|---|
| (a) Ferromagnetic substance is strongly attracted by magnet. | (a) Feebly attracted by magnet. | (a) Feebly repelled by magnet. |
| (b) Retains magnetism when magnetising force is removed. | (b) Does not. | (b) Does not. |
| (c) Susecptibility $\chi >> 1$. | (c) $\chi$ is small but positive. | (c) $\chi$ is small but negative. |
| (d) Permeability $\mu >> 1$. | (d) $\mu$ is slightly greater than 1. | (d) $\mu < 1$. |
| (e) Hysteris is a distinguishing property. | (e) No hysterisis. | (e) No hysterisis. |
| (f) Above Curie point ferromagnetic substance becomes paramagnetic. | (f) At a very high temperature paramagnetic substances behave like diamagnetic substances. | (f) Temperature has no effect on diamagnetic substances. |
| (g) Ferromagnetic substances are crystalline solids. | (g) Paramagnetic substances exist in all three states, viz., solid, liquid and gas. | (g) Diamagnetic substances exist in all three states. |

## SOLVED PROBLEMS

**1.** A ferromagnetic material with $J = \frac{3}{2}$ and $g = 2$ has a Curie temperature of 120 °K. Calculate the intrinsic flux density near 0 °K. Also calculate the ratio of the magnetisation at 290 °K in the presence of an external field of mT to the spontaneous magnetization at 0 °K.

**Soln.** From equation (5.53), we can write,

$$B_E = \lambda M$$

But from equation (5.59), we can write,

$$M_s(0) = Ng\, J\mu_B$$

$$\therefore \quad B_E = \lambda NgJ\mu_B$$

Using equation (5.67), we get

$$B_E = \frac{3K\,\theta_c}{g\mu_B(J+1)}$$

$$= \frac{3 \times 1.38 \times 10^{-23} \times 120}{2 \times 9.27 \times 10^{-24} \times \left(\frac{3}{2}+1\right)}$$

$$= \mathbf{107.187} \text{ tesla}$$

The magnetization for $T > \theta_C$ is obtained from equation (5.65) as

$$M = \frac{CB}{\mu_0 (T - \theta_c)}.$$

Using the equations (5.66) and (5.67), we get

$$M = \frac{Ng^2\mu_B^2 J(J+1)B}{3K(T-\theta_c)}.$$

But from equation (5.59),

$$M_s(0) = Ng\, J\mu_B.$$

$$\therefore \quad \frac{M}{M_S(0)} = \frac{g\mu_B(J+1)B}{3K(T-\theta_c)}$$

$$= \frac{2\times 9.27\times 10^{-24}\times \frac{5}{2}\times 10^{-3}}{3\times 1.38\times 10^{-23}\times (300-120)}$$

$$= \mathbf{3.62 \times 10^{-5}}.$$

2. The magnetic field strength in a piece of metal is $10^6$ amperes per metre. Find the flux density and the magnetisation in the metal. Assume that the magnetic susceptibility of the metal is $1.4 \times 10^{-3}$.

**Soln.** We can write from equation (5.1)

$$\chi = \frac{M}{H}$$

or,

$$M = \chi H$$
$$= 1.4 \times 10^{-3} \times 10^6$$
$$= \mathbf{1.4 \times 10^3\ A/m}.$$

Again, from equation (5.2), we have

$$B = \mu_0 (H + M)$$
$$= 4\pi \times 10^{-7} (10^6 + 1.4 \times 10^3)$$
$$= \mathbf{1.2584\ tesla}.$$

## REVIEW QUESTIONS

1. Define magnetization and magnetic susceptibility. Write down relationship between magnetic flux density, the magnetic intensity $H$ and the magnetization $M$.
2. (a) Give an account of quantum theory of paramagnetism and derive an expression for susceptibility. **[Luck. Univ. 1992]**
   (b) What is Curie's law? Name two paramagnetic elements. What is Pauli Paramagnetism? **[C.U. 2006]**
3. Write a note on (a) Antiferromagnetism (b) Ferrimagnetism.
4. Describe the Langevin's theory of diamagnetism and obtain an expression for diamagnetic susceptibility.

5. What is ferromagnetism? Discuss the Weiss theory of ferromagnetism and obtain an expression for ferromagnetic susceptibility.

6. Find an expression of paramagnetic susceptibility from quantum theory. Under what condition does it deduce to the classical expression of Langevin? **[C.U. 2005]**

7. (a) What are the characteristics of diamagnetic, paramagnetic and ferromagnetic substances? How would you distinguish between the three types of substances experimentally? **[C.U. 1995]**

   (b) Give a theory which explains why ferromagnetic substance becomes paramagnetic above a critical temperature? **[C.U. 1995]**

8. (a) Discuss the characteristics of diamagnetic, paramagnetic and ferromagnetic substances. How would you distinguish between the three types of substances experimentally? **[C.U. 1997]**

   (b) (i) Name an element which exhibits paramagnetism. **[C.U. 1997]**

   (ii) Explain why the inert gases do not exhibit paramagnetism.

   (c) Derive Curie-Weiss law of ferromagnetic susceptibility and compare the temperature dependence of magnetization obtained from it with the experimental results. **[C.U. 1999]**

9. Give an account of the quantum theory of paramagnetism and discuss the low and high temperature cases. What is Curie's law? Name two paramagnetic elements. What is Pauli paramagnetism?

10. At about 180 °K iron transforms into a FCC structure from a BCC structure. Which is the structural form at the room temperature? Assuming no change in density, find the ratio of nearest neighbour distance in FCC structure to that in the BCC structure. **[C.U. 2007]**

11. (a) State the basic assumptions of the molecular field theory of ferromagnetism.

    (b) Derive the Curie-Weiss law of ferromagnetism.

    (c) Draw the susceptibility versus temperature curve, for temperatures above the Curie temperaure. **[C.U. 2007]**

12. (a) Distinguish between paramagnetic and ferromagnetic substances? Give examples. **[C.U. 2007]**

    (b) Discuss Langevin theory for a paramagnetic gas and obtain an expression for paramagnetic susceptibilities of the gas. **[C.U. 2008]**

## PROBLEMS

1. Calculate the diamagnetic susceptibility of copper assuming that an atom of copper contributes only one electron and mean square distance of electronic distribution from the nucleus is equal to the radius of monovalent copper ion. The lattice constant of copper is 3.61 Å and the radius of $Cu^+$ is 0.94 Å. **[Ans.** $-4.62 \times 10^{-6}$**]**

2. Iron is a ferromagnet with Curie temperature of 1043 °K and an effective moment of $2.2\mu_B$ per atom. Compute the internal field for iron. **[Ans.** 2100 T**]**

3. The Curie temperature of iron is 1043 °K. If each iron atom has a magnetic moment of two Bohr magneton, calculate the values of the Curie constant. Assume that the saturation magnetization of iron is $1.75 \times 10^6$ ampere/metre. **[Ans.** 9848 K**]**

# 6
# Semiconductor

## 6.0 INTRODUCTION

Semiconductors are material whose electrical properties are intermediate between those of metals (good conductor) and insulators. The electrical conductivity of semiconductor increases with the rise of temperature, i.e., it has *negative temperature coefficient* of resistance. *Germanium* and *Silicon* are most common semiconductors which belong to Group IV of the periodic table. There are certain compound semiconductors such as Gallium Arsenide (GaAs), Indium Phosphid (InP), Cadmium Sulphide (CdS), etc. Semiconductors have some useful properties that are extensively used in electronic circuit. Particularly utilization of semiconductors in semiconductor diodes, radios, solar batteries, etc. is fast replacing bulky vacuum tube devices. *Transistor* development by J. Bardeen, W. Brattain and W. Shockley in 1948 is the climax of investigation. Semiconductors are classified as follows:

(i) Pure or intrinsic semiconductor

(ii) Impurity or extrinsic semiconductor.

A semiconductor in an extremely pure form is known as pure or *intrinsic semiconductor*. The process of introduction of small amount of suitable impurity in a semiconductor is known as *doping*. It is then called impurity or *extrinsic semiconductor*.

## 6.1 INTRINSIC SEMICONDUCTOR

A semiconductor in an extremely pure form is known as an *intrinsic semiconductor* such as pure Germanium or Silicon. Electrical conductivity of this type of semiconductor is solely determined by thermally generated carriers. To understand the mechanism of conduction consider the case of Gernanium which has atomic number 32. Each Germanium atom has four valence electrons and can form four covalent bonds with neighbouring Germanium atom. The Germanium crystal exhibits a three-dimensional regular network type structure which for simplicity is represented by a two-dimensional network and is shown in Figure 6.1. At low temperature (0 °K) all valence electrons in Germanium crystal form covalent bond and no free electron exists and Germanium then behaves like pure insulator. As temperature rises electrons move at random in the crystal by breaking the covalent bond. Each escaped electron leaves behind an empty space called a *hole* (Figure 6.2 and Figure 6.3). *Hole* is positively charged and moves towards the negative terminal of the supply.

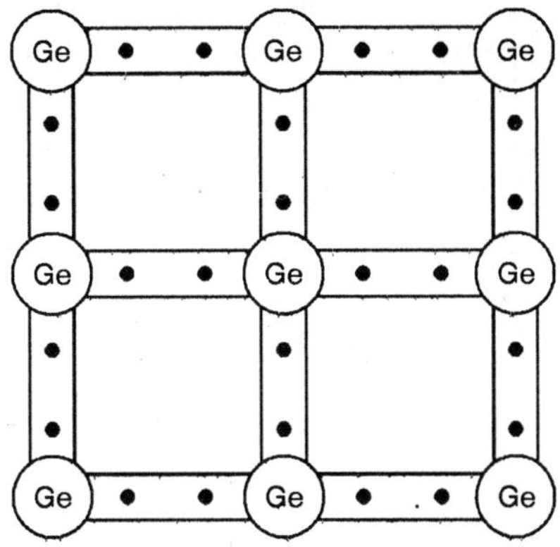


**Figure 6.1** A two-dimensional network of Germanium crystal

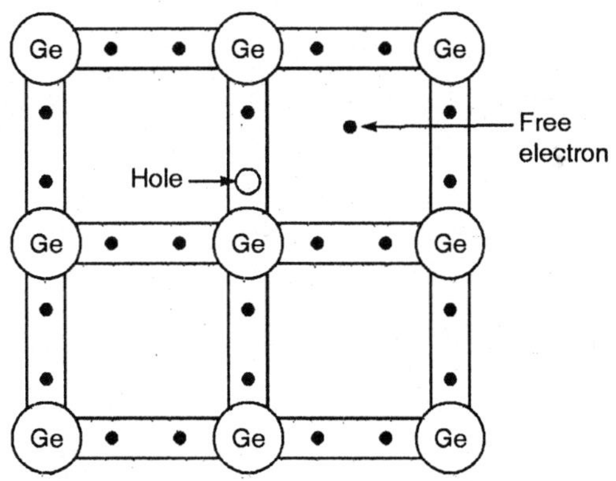


**Figure 6.2** Creation of an electron-hole pair by the rupture of a covalent bond

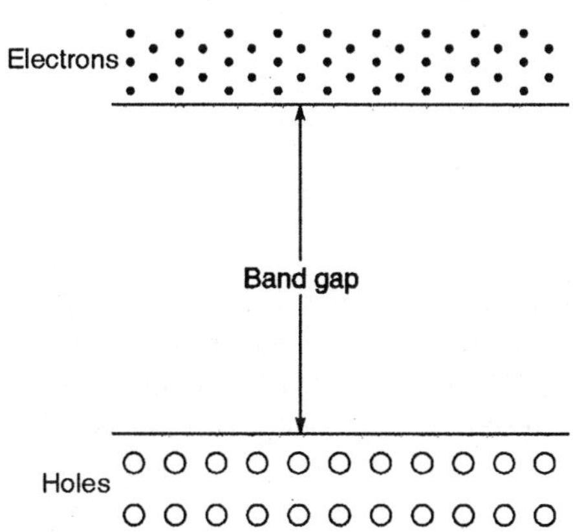


**Figure 6.3** Electrons in conduction band and hole in valence band with band gap in a semiconductor

These electrons and holes move in opposite directions under the effect of an external field and constitute the current. In an intrinsic semiconductor in equilbrium the number of holes is equal to the number of free electrons. If $n_i$ be the number of electrons and $p_i$ the number of holes in the valence band, and if $n_i = p_i = n$, $n$ is called the *intrinsic carrier concentration*.

## 6.2 EXTRINSIC OR IMPURITY SEMICONDUCTOR

The conductivity of intrinsic semiconductor is very small at room temperature. The introduction of small amount of suitable impurity to intrinsic semiconductor increases the conductivity of the intrinsic semiconductor. It is then called impurity or *extrinsic semiconductor.* The process of adding of impurity to a semiconductor in controlled manner is called *doping*. Depending on the type of doping excess free electrons or holes are generated in the semiconductor. In semiconductor, if the dopant is an element of Group V in periodic table excess free electrons in Germanium crystal will be formed and make the semiconductor *n*-type or donor type semiconductor (Figure 6.4). Group V elements in

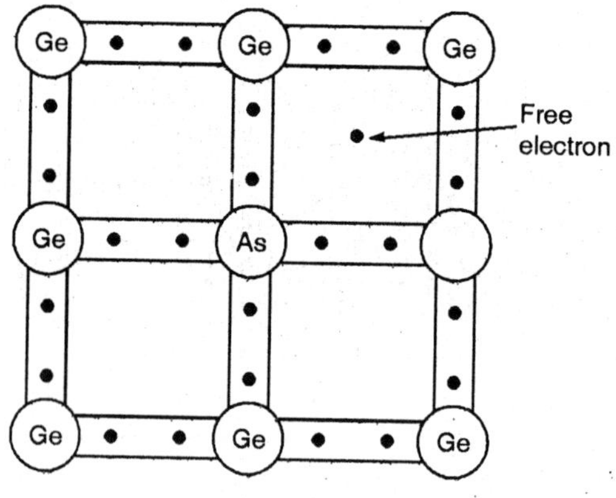


**Figure 6.4** Crystal lattice with a Ge atom substituted by pentavalent As atom

the periodic table are Arsenic, Antimony, Phosphorus, etc. On the other hand, if the dopant is an element of Group III in the periodic table excess holes are created in Germanium and make the semiconductor *p*-type or acceptor type semiconductor (Figure 6.5). Boron, Gallium, Indium, etc.

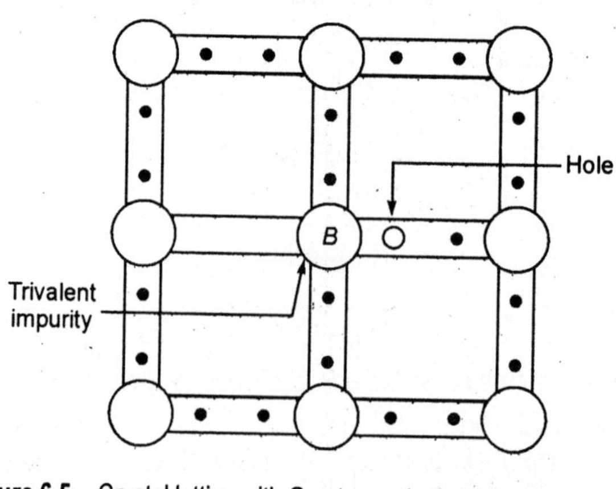


**Figure 6.5** Crystal lattice with Ge atom substituted by trivalent *B* atom

are common Group III elements in the *periodic table* having three valence electrons. Depending on the type of impurity added extrinsic semiconductors are classified as

(a) *n-type* or *donor type semiconductor.*

(b) *p-type* or *acceptor type semiconductor.*

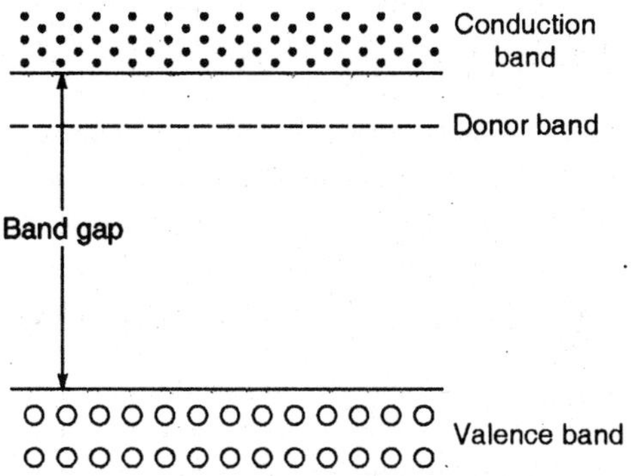


**Figure 6.6** Donor level in the energy band diagram for an *n*-type semiconductor

### (a) *n*-type or donor type semiconductor

When a small amount of pentavalent impurity atom of Group V in the periodic table such as Antimony, Arsenic, Phosphorous, etc. is introduced in a pure semiconductor such as Germanium or Silicon, then *n-type* or *donor type semiconductor* is formed. The size of the impurity atom is similar to Germanium or Silicon atoms in the crystal lattice. Four of the five valence electrons of the pentavalent impurity element such as Arsenic (As) as impurity atom will form covalent bonds with four valence electrons of the neighbouring host crystal, viz., Germanium or Silicon as in Figure 6.4. When impurity atoms As donates excess free electrons, then semiconductors are known as *n*-type impurity or donors. The electrons are negativly charged particles in the semiconductor having donor type impurties which are known as *n-type semiconductors*. The energy level corresponding to fifth valence electron of the impurity atom lies in the band gap just below the conduction band edge as in Figure 6.6. Electrons are the main current carriers in donor type semiconductor and are majority carriers, and thermally generated holes are minority carriers. If $n$ and $p$ are respectively electron concentration, and hole concentration, then $np = n_i^2$, where $n_i$ are intrinsic carrier concentration.

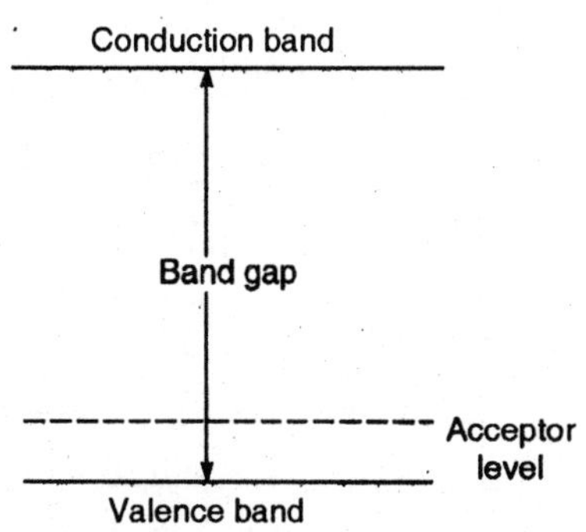


**Figure 6.7** Acceptor level in the energy band diagram for a *p*-type semiconductor

### (b) *p*-type or acceptor type semiconductor

When a small amount of *trivalent* impurity atoms of Group III in the periodic table like Aluminium, Boron, Gallium, etc., is introduced in a pure semiconductor such as Germanium or Silicon, then *p*-type or acceptor type semiconductor is formed. The size of the impurity atom is similar to Germanium or Silicon atoms in the crystal lattice. Three valence electrons of the trivalent Aluminium impurity atom form three covalent bonds with the neighbouring host crystal, viz., Ge or Si as in Figure 6.5, while fourth bond is not completed due to the deficiency of one electron. The missing electron is called hole. The current conduction in *p*-type or acceptor type semiconductor is predominant by holes, i.e., positive charge. So the holes are the majority carrier in *p*-type or acceptor type semiconductor and the thermally generated free electrons are minority carriers in this type of semiconductor. The energy level corresponding to *p*-type impurities lie in the band gap just above valence band as in Figure 6.7. In this type of semiconductor hole concentration $p$ is quite large as compared to electron concentration $n_p$, then $n_p = n_i^2$, where $n_i$ is the intrinsic carrier concentration.

## 6.3 DRIFT VELOCITY, ELECTRICAL CONDUCTIVITY AND CARRIER MOBILITY

At ordinary temperature an intrinsic semiconductor contains some number of free electrons in conduction band and an equal number of holes in the valence band. In the absence of the electric field these electrons and holes move at random and do not contribute any current. It is evident that by the influence of electric field electrons and holes get accelerated towards the opposite end of the field and velocity begins to increase. This increase in velocity becomes steady with the collisions of these electrons and holes with atomic nuclei, phonons, etc. present in the semiconductor. The extra velocity acquired by electrons and holes in the influence of applied field is called the *drift velocity*. The drift velocity is denoted by $v$ and is proportional to the field strength $E$ of the applied electric field, i.e.,

$$v \propto E$$

or,
$$v = \mu E \tag{6.1}$$

or,
$$\mu = \frac{v}{E}$$

where $\mu$ is defined as the *mobility of the charge carrier*. The mobility of charge carrier is defined as the magnitude of the drift velocity for unit electric field. The mobility is defined as positive for both electrons and holes even though their drift velocities are in opposite direction. The S.I. unit of mobility is m²/V-s. In an ideal instrinsic semiconductor mobility depends on lattice scattering. The lattice scattering is dominant at higher temperature, and at lower temperature seattering by electrons may dominate.

The current density, i.e., the current through unit cross-sectional area for electrons, and holes is given by

$$J_e = n_e e v \tag{6.2}$$

$$J_h = n_h e v, \tag{6.3}$$

where $e$ is the charge of the electron, $n_e$ the concentration of electrons in the conduction band and $n_h$ the concentration of holes in the valence band.

With the help of equation (6.1) we can write the equations (6.2) and (6.3) as

$$J_e = n_e \, e \, \mu_e \, E \tag{6.4}$$

$$J_h = n_h \, e \, \mu_h \, E. \tag{6.5}$$

But we know from the Ohm's law,

$$J_e = \sigma_n \, E \tag{6.6}$$

and
$$J_h = \sigma_h \, E, \tag{6.7}$$

where $\sigma_n$ and $\sigma_h$ are the conductivities of electrons and holes respectively.

Comparing equations (6.4) and (6.5) with the equations (6.6) and (6.7) we can write,

$$\sigma_n = n_e e\, \mu_e \tag{6.8}$$

$$\sigma_h = n_h e\, \mu_h. \tag{6.9}$$

The electrical conductivity in the presence of both electrons and holes is given by the sum of the separate contribution:

$$\sigma = \sigma_n + \sigma_h = n_e e \mu_e + n_h e \mu_h. \tag{6.10}$$

For intrinsic semiconductor,

$$n_e = n_h = n.$$

Therefore, the equation (6.10) becomes

$$\sigma = ne\,(\mu_e + \mu_h). \tag{6.11}$$

For $n$-type semiconductor $n_e >> n_h$, so the equation (6.11) becomes

$$\sigma_{n\text{-}type} = ne\mu_e. \tag{6.12}$$

Similarly, for $p$-type semiconductor $n_h >> n_e$, so the equation (6.11) is

$$\sigma_{p\text{-}type} = ne\mu_h. \tag{6.13}$$

## 6.4 CARRIER CONCENTRATION AND FERMI LEVELS IN INTRINSIC SEMICONDUCTOR

The conductivity of a semiconductor is determined by the concentration of electrons and holes in a semiconductor which can be obtained from the density of available states in valence band and conduction band and Fermi-Dirac distribution function.

The number of electrons per unit volume occupying states in the energy range between $E$ and $E + dE$ in any electronic system in thermal equilibrium as in equation (3.67) is

$$dN = N(E)\, dE = D(E)\, F(E)\, dE, \tag{6.14}$$

where $F(E)$ is the Fermi distribution function and is given in equation (3.62) as

$$F(E) = \frac{1}{e^{(E-E_F)/KT} + 1} \tag{6.15}$$

and the density states $D(E)$ is defined as the number of electronic states per unit volume in a semiconductor and is given by the equation (3.60) as

$$D(E) = \frac{1}{2\pi^2}\left(\frac{2m}{\hbar^2}\right)^{3/2} E^{1/2}$$

$$= \frac{4\pi}{h^3}(2m)^{3/2} E^{1/2}. \tag{6.16}$$

Using equations (6.14) and (6.16), we obtain

$$dN = \frac{4\pi}{h^3}(2m)^{3/2} E^{1/2}\, F(E)\, dE. \tag{6.17}$$

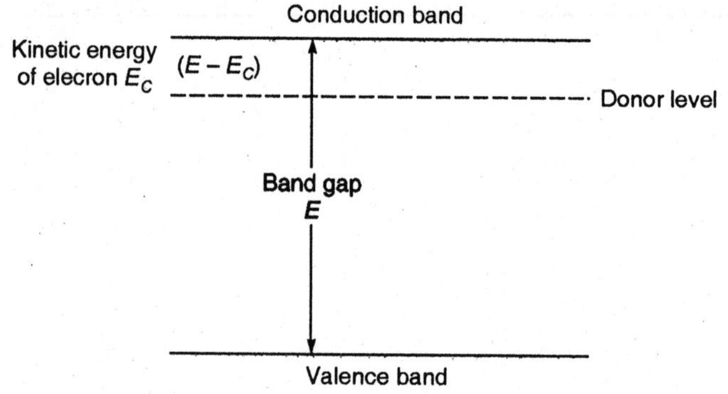


**Figure 6.8** Band model of a semiconductor

An electron occupying an energy state $E$ in the *conduction band* actually possesses a kinetic energy $(E - E_C)$ as in Figure 6.8. Therefore, equation (6.17) with the help of equation (6.15) becomes

$$dN = \frac{4\pi}{h^3}\left(2m^*\right)^{3/2}\left(E - E_C\right)^{1/2}\frac{1}{e^{(E-E_F)/KT}+1}\,dE, \tag{6.18}$$

where $m^*$ is the effective mass of the electron in the *conduction band* as because the electron in the conduction band is not free but interact with the periodic potential of the semiconductor crystal, so we substitue effective mass of the electron $m_e^*$ instead of $m^*$. Hence, the density of electon in the conduction band is

$$N = \frac{4\pi}{h^3}\left(2m_e^*\right)^{3/2}\int_{E_F}^{\infty}\frac{(E - E_C)^{1/2}\,dE}{e^{(E-E_F)/KT}+1}. \tag{6.19}$$

Near room temperature $KT$ has very small value, so $(E - E_F) >> KT$, so $e^{(E-E_F)/KT} >> 1$. So the equation (6.19) becomes

$$N = \frac{4\pi}{h^3}\left(2m_e^*\right)^{3/2}\int_{E_C}^{\infty}\frac{(E - E_C)^{1/2}\,dE}{e^{(E-E_F)/KT}}$$

$$= \frac{4\pi}{h^3}\left(2m_e^*\right)^{3/2}\int_{E_C}^{\infty}(E - E_C)^{1/2}\,e^{-(E-E_F)/KT}\,.dE$$

$$= \frac{4\pi}{h^3}\left(2m_e^*\right)^{3/2}\int_{E_C}^{\infty}(E - E_C)^{1/2}\,e^{-(E-E_C)/KT}\,.dE$$

$$= e^{(E_F-E_C)/KT}\,dE\,.$$

Put $\frac{E - E_C}{KT} = x$.

$\therefore \quad dE = KT\,dx$

and for $E = E_C$, $x = 0$.

$$\therefore \quad N = \frac{4\pi}{h^3}\left(2m_e^*\right)^{3/2} E^{(E_F E_C)/KT}\int_0^\infty (KT)^{1/2} x^{1/2} e^{-x}.KT.dx$$

$$= \frac{4\pi}{h^3}\left(2m_e^* KT\right)^{3/2} e^{(E_F - E_C)/KT} \times \int_0^\infty x^{1/2} e^{-x} dx \tag{6.20}$$

$$= \frac{4\pi}{h^3}\left(2m_e^* KT\right)^{3/2} e^{(E_F - E_C)/KT}\left(\frac{\pi}{4}\right)^{1/2}.$$

$$\because \quad \left[\int_0^\infty x^{1/2} e^{-x} dx = \left(\frac{\pi}{4}\right)^{1/2}\right],$$

$$\therefore \quad N = 2\left(\frac{2\pi\, m_e^* KT}{h^2}\right)^{3/2} e^{(E_F - E_C)/KT}. \tag{6.21}$$

The equation (6.21) gives the concentration of electrons in conduction band.

**Concentration of holes in the valence band**

Hole is created when an electron from the valence band jumps to conduction band. In equilibrium the density of the holes in the energy range $E$ and $E + dE$ is given by

$$N_h = \int_{-\infty}^{E_v} D(E)\left[1 - F(E)\right] dE\,. \tag{6.22}$$

We have replaced Fermi function $F(E)$ in equation (6.15) by $[1 - F(E)]$ which represents the probability that it may be unoccupied by an electron, i.e., the probability of finding hole in the energy state $E$. In equation (6.22), $E_\upsilon$ is the energy at the top of the valence band. With the help of equation (6.15) we can write,

$$1 - F(E) = 1 - \frac{1}{e^{(E-E_F)/KT} + 1}$$

$$= \frac{e^{(E-E_F)/KT}}{e^{(E-E_F)/KT} + 1}. \tag{6.23}$$

In the valence band, $E << E_F$, $e^{(E-E_F)/KT} << 1$. Therefore, the equation (6.23) can be written as

$$1 - F(E) = e^{(E-E_F)/KT}. \tag{6.24}$$

Let $E_\upsilon$ be the energy at the top of the valence band, then the kinetic energy of a hole in the energy state $E$ in the valence band is $(E_V - E)$. So the density of states per unit volume is

$$D(E) = \frac{4\pi}{h^3}\left(2m_h^*\right)^{1/2}\left(E_V - E\right)^{1/2}, \tag{6.25}$$

where $m_h^*$ is the effective mass of a hole in the valence band. The hole concentration in the valence band is obtained by using equations (6.24), (6.25) in equation (6.22) and integrating from $E = -\infty$ to $E = E_v$ as

$$N_h = \frac{4\pi}{h^3}\left(2m_h^*\right)^{3/2} e^{(E_V - E)/KT} \int_{-\infty}^{E_V} \left(E_V - E\right)^{1/2} e \frac{(E - E_V)}{KT} .dE \,. \tag{6.26}$$

To solve the equation (6.26), we put

$$\frac{E_v - E}{KT} = x$$

or, $$dE = -KTdx$$

and when $E = E_V$, $x = 0$ and when $E = -\infty$, $x = \infty$.

The equation (6.26) becomes

$$N_h = \frac{4\pi}{h^3}\left(2m_h^*\right)^{3/2} e^{(E_V - E_F)/KT} \int_{\infty}^{0} x^{1/2} e^{-x} (KT)^{1/2} (-KT)\, dx$$

$$= \frac{4\pi}{h^3}\left(2m_h^* KT\right)^{3/2} e^{(E_V - E_F)/KT} \int_{0}^{\infty} x^{1/2} e^{-x} dx$$

$$= \frac{4\pi}{h^3}\left(2m_h^* KT\right)^{3/2} e^{(E_V - E_F)/KT} \left(\frac{\pi}{4}\right)^{1/2}$$

$$= 2\left(\frac{2\pi m_h^* KT}{\hbar^2}\right)^{3/2} e^{(E_V - E_F)/KT}$$

or, $$N_h = N_v\, e^{-(E_F - E_V)/KT}, \tag{6.27}$$

where $N_V = 2\left(\frac{2\pi m_h^* KT}{h^2}\right)^{3/2}$ is the effective density of the holes at the valence band.

In the case of an intrinsic semiconductor the electron and hole concentration are equal, i.e., $N = N_h$, then from equations (6.21) and (6.27) we can write,

$$2\left(\frac{2\pi m_e^* KT}{h^2}\right)^{3/2} e^{(E_F - E_C)/KT} = 2\left(\frac{2\pi m_h^* KT}{h^2}\right)^{3/2} e^{(E_V - E_F)/KT} . \tag{6.28}$$

Simplifying equation (6.28), we get

$$E_F = \frac{E_V - E_C}{2} - \frac{3}{4} KT \,\text{In}\, \frac{m_e^*}{m_h^*} . \tag{6.29}$$

The equation (6.29) gives the position of the Fermi level in a pure semiconductor. If $m_e^* = m_h^*$, then the equation (6.29) becomes

$$E_F = \frac{E_V - E_C}{2}. \tag{6.30}$$

So the Fermi level lies in the middle of the forbidden gap as in Figure 6.9.

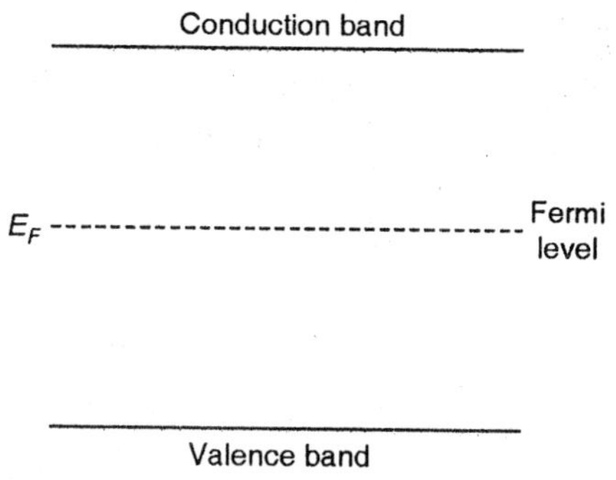


**Figure 6.9** Fermi level lies in the middle of the conduction band and valence band

In general, $m_e^* < m_h^*$ , so $E_F$ is closer to the conduction band and rises slowly as temperature $T$ increases.

The intrinsic carrier concentration can be obtained from equations (6.21) and (6.27) as

$$N_i^2 = NN_h = 4\left(\frac{2\pi\, KT}{h^2}\right)^3 \left(m_e^* m_h^*\right)^{3/2} e^{-Eg/KT}, \tag{6.31}$$

where $E_g = E_C - E_V$ is the energy band gap. The smaller the energy band gap and higher the temperature, larger is the intrinsic carrier concentration. We find from equation (6.31) that at a given temperature the product of the concentration of electrons in the conduction band and that of holes in the valence band of a semiconductor is independent of Fermi level but depends on the energy band gap $E_g = E_C - E_V$. For intrinsic semiconductor, $N = N_h = N_i$.

$$\therefore \quad N_i = \left(NN_h\right)^{1/2} = 2\left(\frac{2\pi KT}{h^2}\right)^{3/2} \left(m_e^* m_h^*\right)^{3/4} e^{-Eg/2KT}. \tag{6.32}$$

The equation (6.32) gives the value of the concentration of electrons in the conduction band and the value of the holes in the valence band of a semiconductor in terms of the energy gap.

## 6.5 MODEL FOR AN IMPURITY SEMICONDUCTOR

Most semiconductors owe their conductivity due to the presence of foreign atoms built into the lattice or to a stoichiometric excess of one of its constituents. At absolute zero such a solid may contain a certain concentration of the occupied electronic levels which lie in the normally forbidden region between valence and conduction bands. When electrons from these levels are excited into the vacant conduction band the substance is $n$-type semiconductor. The impurity level is called *donor levels.* Conversely, if the impurity semiconductor has certain density of holes at absolute zero, it is trapped in levels lying in the forbidden gap which accepts electrons from the filled band when temperature is slightly increased. These excited electrons leave a hole in filled valence band and conduction becomes possible in this band. This is *p*-type semiconductor. The donor and acceptor states are shown in Figures 610(a) and 6.10(b).

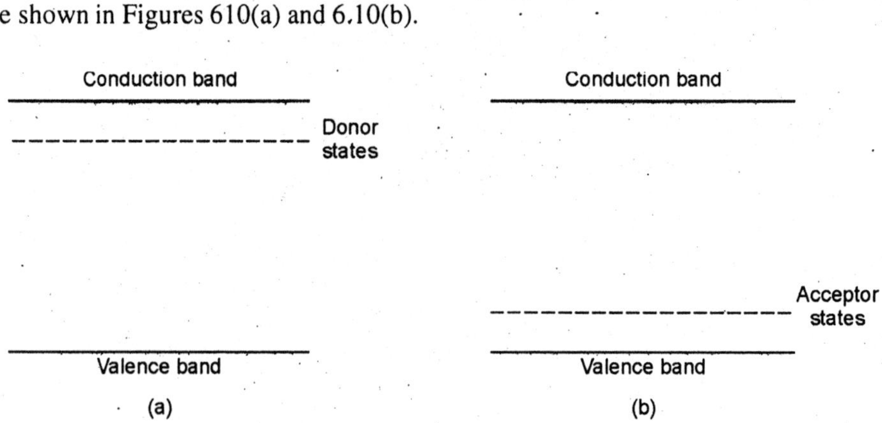


**Figure 6.10** (a) Energy level diagram for *n*-type semiconductor, (b) Energy level diagram for *p*-type semiconductor

### (a) *n*-type semiconductor

An *n*-type semiconductor consists of a donor level of Group V element just below the conduction band as in Figure 6.10(a). Let the donor level have energy $E_d$, and $N_d$ the concentration of the ionised donor. As the donor level lies close to the conduction band we can neglect the effect of valence band for the moment. Therefore, we may suppose that the concentration of electrons in the conduction band is equal to the concentration of vacancy in the donor level. Let us assume that at absolute zero all the donor levels are filled with electrons. At low temperature when only a small fraction of donors are ionised, we expect Fermi level $E_F$ to lie about halfway between the donor levels and the bottom of the conduction band. The density of electrons in the conduction band is given by equation (6.21) as

$$N = 2\left(\frac{2\pi m_e^* KT}{h^2}\right)^{3/2} e^{(E_F - E_C)/2KT},$$

where $E_F$ and $E_C$ are the energy of the Fermi level and lower edge of the conduction band.

If $E_F$ lies more than a few $KT$ above the donor level, the number of vacancies per unit volume of empty donor is given by

$$N = N_d\left[1 - F(E)\right] = N_d\left[1 - \frac{1}{e^{(E_d - E_F)/KT} + 1}\right]$$

$$= N_d \frac{e^{(E_d - E_F)/KT}}{e^{(E_d - E_F)/KT} + 1}$$

$$= N_d \frac{1}{1 + e^{(E_F - E_d)/KT}}. \qquad (6.33)$$

For $T \to 0$ the unit term in the denominator of equation (6.33) can be neglected, so the equation (6.33) becomes

$$N_d\left[1-F(E)\right]=N_d\, e^{(E_d-E_F)/KT}.$$

Now equating the equations (6.21) and (6.32), we obtain

$$2.\left(\frac{2\pi m_e^* KT}{h^2}\right)^{3/2} e^{(E_F-E_C)/KT}=N_d\, e^{(E_d-E_F)/KT}. \qquad (6.34)$$

Taking natural logarithm on both sides of equation (6.33), we have

$$\ln 2\left(\frac{2\pi m_e^* KT}{h^2}\right)^{3/2}+(E_F-E_C)/KT=\ln N_d+(E_d-E_F)/KT$$

or,

$$\left[2E_F-(E_C+E_d)\right]/KT=\ln N_d-\ln 2\left(\frac{2\pi m_e^* KT}{h^2}\right)^{3/2}$$

or,

$$2E_F-(E_C+E_d)=KT\ln\frac{N_d}{2\left(\frac{2\pi m_e^* KT}{h^2}\right)^{3/2}}$$

or,

$$E_F=\frac{E_C+E_d}{2}+\frac{KT}{2}\ln\frac{N_d}{2\left(\frac{2\pi m_e^* KT}{h^2}\right)^{3/2}}. \qquad (6.35)$$

Thus at $T = 0$ °K, second term in equation (6.35) becomes zero and we have

$$E_F=\frac{E_C+E_d}{2}.$$

Thus Fermi level $E_F$ lies exactly halfway between donor level and the bottom of the conduction band as in Figure 6.11. As $T$ increases, Fermi level drops and at room temperature it is below the donor level.

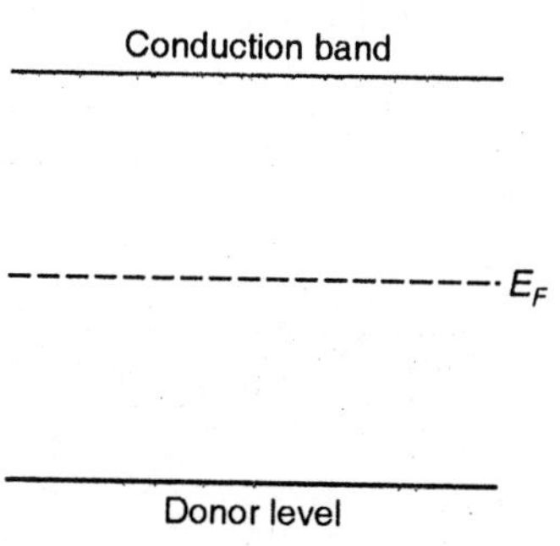


**Figure 6.11** Energy level diagram for *n*-type semiconductor

At room temperature and above, almost all the donor atoms are ionised. At this stage the number of vacancies in donor level can be taken to be $N_d$ and as $(E_d - E_F) >> KT$, hence from equation (6.35), we can write,

$$N_d = 2\left(\frac{2\pi m_e^* KT}{h^2}\right)^{3/2} e^{(E_F - E_C)/KT} \tag{6.36}$$

or,

$$\frac{N_d}{2\left(\frac{2\pi m_e^* KT}{h^2}\right)^{3/2}} = e^{(E_F - E_C)/KT}$$

or,

$$E_F = E_C + KT \ \text{In} \frac{N_d}{2\left(\frac{2\pi m_e^* KT}{h^2}\right)^{3/2}} \tag{6.37}$$

The equation (6.37) gives the position of the Fermi level at room temperature and above.

**(b) *p*-type semiconductor**

When a small amount of trivalent impurity is added to a pure semiconductor, it is called *p*-type semiconductor. In the *p*-type semiconductor an acceptor level lies just above the top of the valence band. Let us suppose that the acceptor level has an energy $E_a$, and $N_a$ is acceptor atoms per unit volume. Since the acceptor level lies close to the valence band we shall neglect the effect of conduction band.

The concentration of ionized acceptor is given by

$$N_a F(E) = \frac{N_a}{e^{(E_a - E_F)/KT} + 1}. \tag{6.38}$$

Since $(E_a - E_F)$ is large compared to $KT$, we can write the equation (6.38) as

$$\frac{N_a}{e^{(E_a - E_F)/KT}} = N_a \, e^{(E_F - E_a)/KT}. \tag{6.39}$$

The hole concentration in the valence band is given by equation (6.27) as

$$N_h = 2\left(\frac{2\pi m_h^* KT}{h^2}\right)^{3/2} e^{(E_V - E_F)/KT}.$$

If the effect of conduction band has been neglected, then the number of holes in the valence band is given from equation (6.38) as

$$N_h = 2\left(\frac{2\pi m_h^* KT}{h^2}\right)^{3/2} e^{(E_V - E_F)/KT} = N_a \, e^{(E_F - E_a)/KT}. \tag{6.40}$$

Taking natural logarithm of both sides of equation (6.40) and simplifying, we have

$$E_F = \frac{E_a + E_V}{2} - KT \ln \frac{N_a}{2\left(\frac{2\pi m_h^* KT}{h^2}\right)^{3/2}}. \tag{6.41}$$

At $T = 0$, the equation (6.41) becomes

$$E_F = \frac{E_a + E_V}{2},$$

i.e., the Fermi level lies exactly halfway between the top of the valence band and acceptor level.

## SOLVED PROBLEMS

1. The intrinsic resistivity of Germanium at 27 °C is 4700 ohm-m. What is the intrinsic carrier concentration when the electron and hole mobilities in Germanium at 27 °C are 0.39 $m^2$/volt-sec and 0.19 $m^2$/volt-sec respectively?

   **Soln.** From equation (6.11), we know the conductivity $\sigma$ is given by

   $$\sigma = ne(\mu_e + \mu_h) = \frac{1}{\text{Resistivity}}.$$

   Here $\mu_e$ at 27 °C = 0.39 $m^2$/volt-sec

   $\mu_h$ at 27 °C = 0.19 $m^2$/volt-sec

   and we know, $e = 1.6 \times 10^{-19}$ coulomb.

   From equation (6.11), we get

   $$n.1.6\times10^{-19}.(0.39+0.19) = \frac{1}{4700}$$

   or, $$n = \frac{1}{4700\times1.6\times10^{-19}\times(0.39+0.19)}$$

   $$= \mathbf{2.293 \times 10^{15}\ /m^2}.$$

2. In intrinsic GaAs the electron and hole mobilities are 0.85 $m^2$/volt-sec and 0.04 $m^2$/volt-sec respectively and the corresponding effective masses are 0.068 $m_0$ and 0.5 $m_0$ respectively, where $m_0$ is the rest mass of an electron. Given the energy band gap at 300 °K at 1.43 eV, determine the intrinsic carrier concentration and conductivity.

   **Soln.** The intrinsic carrier concentration is given from equation (6.32) as

   $$N_i = (NN_h)^{1/2} = 2\left(\frac{2\pi KT}{h^2}\right)^{3/2}(m_e^* m_h^*)^{3/4} e^{-E_g/2KT}$$

   In the problem the following data are supplied as

   $$m_e^* = (0.068)\ m_0$$

   $$m_h^* = 0.5\ m_0$$

$$E_g = 1.43 \text{ eV} = 1.43 \times 1.6 \times 10^{-19} \text{ joule}$$

$$T = 300 \text{ °K}$$

and we have used the standard value of constants

$$K = 1.38 \times 10^{-23} \text{ J per degree kelvin}$$

$$h = 6.63 \times 10^{-34} \text{ J-sec}$$

and $$m_0 = 9.1 \times 10^{-31} \text{ kg}$$

Substituting the above constants, equation (6.32) becomes

$$N_i = 2\left(\frac{2\pi \times 1.38 \times 10^{-23} \times 300}{6.63 \times 10^{-34}}\right)^{3/2} \times \left(0.68 \times 9.1 \times 10^{-31} \times .5 \times 9.1 \times 10^{-31}\right)^{3/4} \times \exp\left(-\frac{1.43 \times 1.6 \times 10^{-19}}{2 \times 1.38 \times 10^{-23} \times 300}\right)$$

$$= \mathbf{1.94 \times 10^{-12}\ m^{-3}}.$$

## REVIEW QUESTIONS

1. What do you mean by (i) an intrinsic semiconductor (ii) an extrinsic semiconductor.
2. Explian the concept of hole.
3. Draw the energy level diagram for (i) an $n$-type semiconductor (ii) a $p$-type semiconductor.
4. Define (i) mobility of a charge carrier (ii) drift velocity of a carrier.
5. What are donor and acceptor?
6. Derive an expression for the electron concentration in an intrinsic semiconductor.
7. Derive an expression for the concentration of holes in the valence band of an intrinsic semiconductor.
8. Derive an expression for the density of electrons in the conduction band of an $n$-type semiconductor. Hence, find the Fermi level at absolute zero.
9. For a $p$-type semiconductor, prove

$$E_F = \frac{E_a + E_V}{2} - \frac{KT}{2} In \frac{Na}{\left(\frac{2\pi m_h^* KT}{h^2}\right)^{3/2}},$$

where $E_a$ is the energy in the acceptor level per unit volume, $E_V$ is the energy in the highest level in valence band and $N_a$ acceptor atoms per unit volume and other constants have usual meaning.

## PROBLEMS

1. In electron and hole mobilities in a Si sample are 0.135 $m^2$/V-s. and 0.048 $m^2$/V-s. Determine the conductivity of intrinsic Si at 300 °K if the intrinsic carrier concentration is $1.5 \times 10^{16}$ atoms/$m^3$. The sample is then doped with $10^{23}$ phosphorous atoms/$m^3$. Determine the

equilibrium hole concentration, conductivity and position of the Fermi level relative to the intrinsic level.

[**Ans.** $4.39 \times 10^{-4}$ $(\Omega m)^{-1}$, $2.25 \times 10^{9}$ $m^{-3}$, $21.6 \times 10^{2}$ $(\Omega m)^{-1}$, 0.406 eV]

**2.** What is the donor concentration in $n$-type Germanium of 100 ohm-m resistivity at 300 °K? Electron mobility = 0.39 $m^2$/volt-sec. [**Ans.** $1.60 \times 10^{21}$]

**3.** The intrinsic carrier density of Germanium at 300 °K is $2.4 \times 10^{17}$ $m^{-2}$. Calculate its intrinsic resistivity if the electron and hole mobilities are 0.35 $m^2$/volt-sec and 0.18 $m^2$/volt-sec respectively. [**Ans.** $49 \times 10^{-2}$ ohm-m]

**4.** For an intrinsic semiconductor with a gap width of 1 eV, calculate the position of Fermi level at $T = 0$ °K and at $T = 300$ °K if $m_h^* = 6m_h^*$ where $m_h^*$ and $m_e^*$ are effective masses of hole and electron respectively. Boltzmann constant $K = 1.4 \times 10^{-23}$ J/°K.

[**Ans.** 0.5 eV, 0.4846 eV]

# 7
# *Defects in Crystals*

## 7.0 INTRODUCTION

A perfect crystal is a very useful concept and exhibits a periodic crystal structure. A perfect crystal is an assembly of atoms distributed periodically in space. It can be built up from a unit cell consisting usually of only a few atoms by means of a set of translations. A real crystal differs from this perfect scheme by the defects it contains which modify its properties. A defect in a crystal is any deviation from periodicity in the crystal. All crystals must have some type of lattice defects. Point defects, linear defects, surface defects, volume defects, etc., are the main defects in crystals. The linear defects in crystal are called *dislocation*. The defect may scatter conduction electrons in the metal as a result of which it increases the electrical resistance. Some defects decrease the strength of the crystal. The colour of many crystals arises from imperfections. The luminiscence of a crystal is connected with impurity. The mechanical and plastic properties of solids are due to imperfections. Amorphous solids may contain defects. For instance in ideally bounded amorphous silica all Si atoms have four bonds to atoms and all O atoms have 2 bonds in Si atom. Thus, for example, an O atom with only one Si bond can be considered a defect in silica.

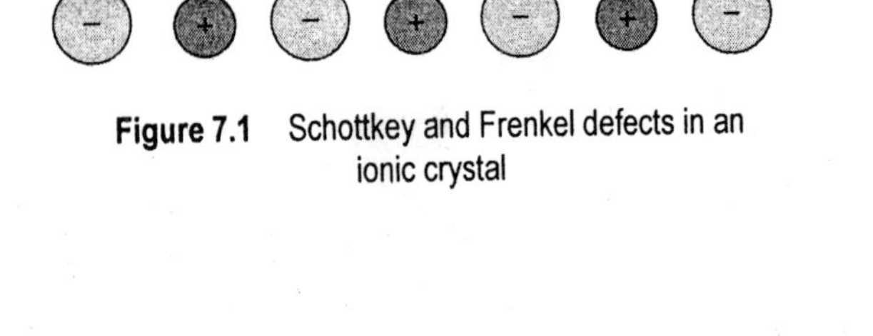


**Figure 7.1** Schottkey and Frenkel defects in an ionic crystal

## 7.1 POINT DEFECTS

Any deviation in a crystal from a perfect periodic lattice is confined to a small region of only about a few lattice constants. It is called *point defects.* These defects have an extension in area and are confined to a small region. A lattice vacancy which is missing in an atom or ion is known as a *Schottky defect*. The common point imperfections are vacant lattice sites, chemical properties and interstitial atoms. When an atom or ion leaves its normal site and is found to occupy another position in interspace, then *Frenkel defects* are created and it is created by interstitial atom and a vacancy (Figure 7.1). Presence of a foreign atom in the lattice in any interstitial position creates also point defects.

## 7.2 DISLOCATION

The linear defects in crystal are called *dislocation.* In the dislocation some atoms of crystal lattice are misaligned. The presence of dislocation results in lattice strain, i.e., distortion. Dislocation has the following criteria:-

1. A dislocation line cannot end in the interior of a crystal. It must close on itself or end at other defect or outer surface grain boundary or node formed with other dislocations.
2. The displacement which defines the dislocation has the same value along its whole length.
3. All displacements can be analysed in terms of translation and rotational components.

There are two types of disolcations:

(i) *Edge dislocation* in which *Burgers vector* is perpendicular to dislocation line (Figure 7.2).

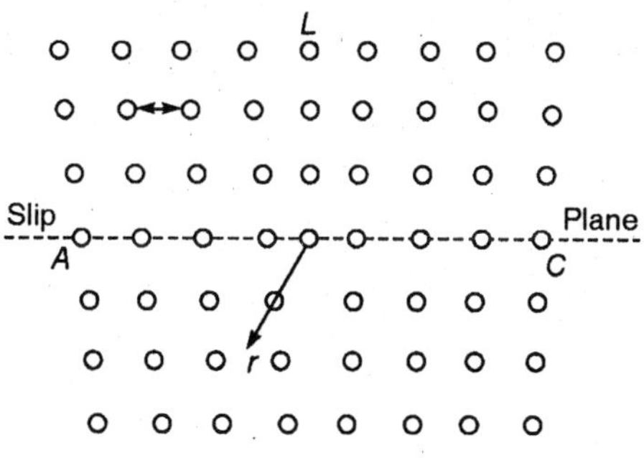


**Figure 7.2** Edge dislocation is shown in the figure

(ii) *Screw dislocation* in which *Burgers vector* is parallel to the dislocation line (Figure 7.3).

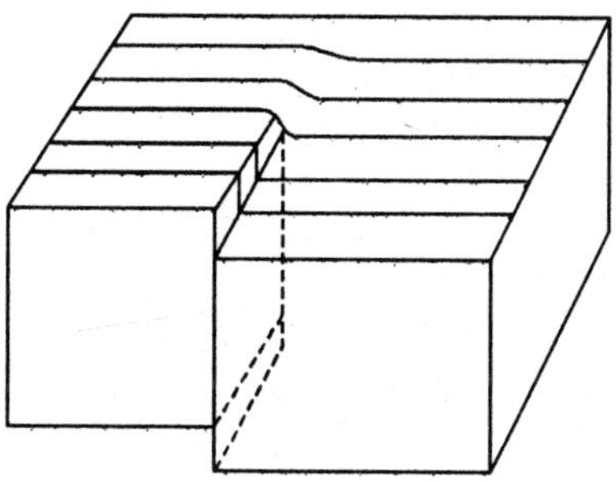

**Figure 7.3** Screw dislocation is shown in the figure

*Burgers vector* $\vec{b}$ denotes actually the dislocation displacement vector or in otherwise displaces the material on one side of a surface relative to other side by a vector $\vec{b}$. The direction and magnitude of such distortion is expressed by Burgers vector $\vec{b}$.

It seems that a good crystal contains about $10^8$ to $10^{10}$ dislocations per square metre distributed on a three-dimensional network. Plastic deformation multiplies these dislocation up to densities in the order of at most $10^{14}$ to $10^{16}$ dislocations per square metre. However, impurities and annealing treatment diminish the dislocation density. Number of dislocation lines per unit area is called *dislocation density.* The presence of dislocations produces strains and stress as in a crystal and therefore, increases the elastic energy stored. The variation of this energy with the position of the dislocations corresponds to an interaction between and with other defects. These dislocations harden the crystal. Dislocation can be observed using transmission electron microscopy, field ion microscopy and atom probe technique. Deep level transient spectroscopy has been used for studying electrical activity of dislocation in semiconductor.

## 7.3 PLANAR DEFECTS

These defects have a change in area and are confined to small region. The notable planar defects are (i) *Grain boundary* (ii) *Stacking fault* (iii) *Linear boundary.* The boundary between two crystals in a poly crystalline solid is called *grain boundary.* This defect occurs due to the geometrical structure of the lattice and is called *static imperfection.* Grain boundaries disrupt the motion of dislocations through the material, so reducing crystal size is a common way to improve strength. Since grain boundaries are defects in crystal structure they tend to decrease the electrical and thermal conductivities of the material. The high interfacial energy and weak boundary in most grain boundaries often make them preferred sites for the onset of corrosion and for the precipitation of new phases from the solid. They are also important to many mechanisms of creep. Grain boundaries may contain special defect only existing in grain boundaries; the most prominent ones are grain boundary dislocations. Grain boundary dislocations are linear defects with all characteristics of lattice dislocations but with very specific Burgers vectors that can occur in grain boundaries.

The plane separating two incorrectly juxtaposed layers is called a *stacking fault.* Lineage boundary is a boundary between two adjacent perfect regions in the same crystal which are slightly tilted with respect to one another.

Other than the above defects there are defects, viz., electronic imperfection and excitation states in crystals.

## REVIEW QUESTIONS

1. What do you mean by defects in crystal?
2. How many types of defects exist in a crystal?
3. Write a note on defects in a crystal.

# 8
# *Superconductivity*

## 8.0 INTRODUCTION

Superconductivity was considered the most extraordinary and mysterious of the properties of metals. Superconductivity was first observed by Kamerlingh Onnes in Leiden in 1911, three years after he first liquified helium. Onnes was investigating the electrical resistance of mercury at a very low temperature below the boiling point of liquid helium and about 4.2 K resistance fell sharply and mercury sample's resistance disappeared below a critical temperature.The change of resistance with temperature is shown in Figure 8.1. This phenomenon of disappearence of electrical resistance in mercurey when cooled for sufficiently low temperature is called *superconductivity*. Many pure metals and metallic alloys exhibit the above phenomenon, i.e., zero resistance at low tempe-ratures. The disappearance of electrical resistance of metals and metallic alloys below a certain temperature was called superconductivity by Kammerlingh Onnes and the materials are known as superconductor. The temperature at which the resistance of a material disappears is known as *critical temperature*. In the critical temperature sample undergoes a phase transition from a normal state to superconducting state (Figure 8.2).

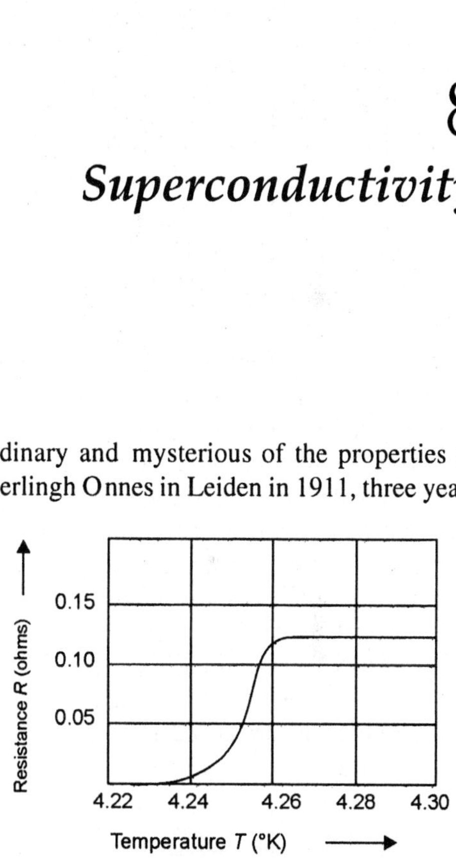


**Figure 8.1** Change of resistance of mercury as a function of temperature

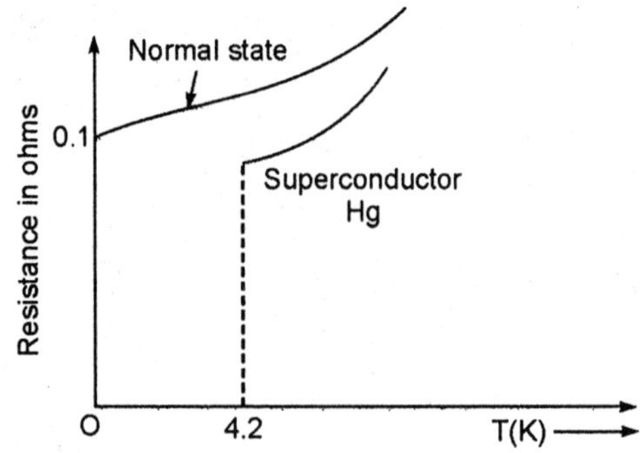


**Figure 8.2** Temperature dependence of the resistance of mercury in normal and superconducting states

No electrical resistance in superconducting state means large electrical current and hence negligible heat losses, i.e., electrical instrument consumes little electical energy in superconducting state. An electromagnet made up of superconducting material can operate for years after years even after removing electrical supply voltage. The disadvantage is that the material is to be operated at a very low temperature, i.e., the liquid helium temperature and to be maintained it for a long time. This disadvantage was recovered by further research for development of material having high critical temperature. The critical temperature of Pb is 7.2 K and Nb is 9.2 K. The critical temperature of $Nb_3Ge$ is 23 K. Critical temperature of La-Ba-CuO system of ceramic superconductor is 34 K. The critical temperature of thallium cuprate is 125 K. The high critical temperature has initiated intense research in superconductivity. Surprisingly those metals which at normal temperature are the best conductors, i.e., Au, Ag and Cu, are not superconductors. Some chemical compounds which are semiconductors at ordinary temperature become superconductor at low temperature. The superconducting properties of metals can be changed by varying (i) temperature, (ii) magnetic field, (iii) magnetic stress, (iv) Impurity, (v) Atomic structure.

## 8.1 EFFECT OF MAGNETIC FIELD ON SUPERCONDUCTOR

Shortly after the observation of disappearance of resistance below critical temperature Kamerlingh Onnes found that superconductivity could be distroyed by the application of very intense magnetic field. The critical magnetic field is defind as magnetic field where half the normal resistance has returned. It is denoted by $H_C$ and is found to be a function of temperature given approximately by Tuyn's law as

$$H_C = H_0\left[1-\left(\frac{T}{T_C}\right)^2\right]. \tag{8.1}$$

where $H_0$ is the critical field at absolute zero ($T_0$).

## 8.2 MEISSNER EFFECT

It is an experimental fact that a bulk superconductor in a weak magnetic field behaves as a perfect diamagnet with zero magnetic induction in the interior of the bulk superconductor. When a superconductor is cooled in a magnetic field below the transition temperature, then at the transition temperature for superconductor the magnetic flux originally present is pushed out of the specimen. This is called *Meissner effect*. This is shown in Figure 8.3. The Meissner effect is a reversible phenomenon.

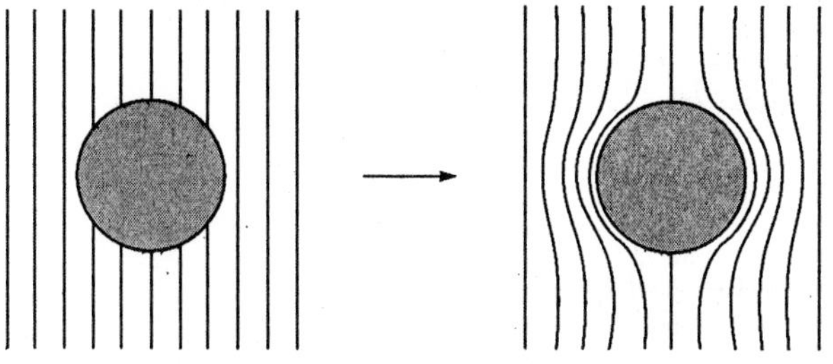

**Figure 8.3(a)** Meissner effect in a superconducting sphere cooled in a constant applied magnetic field on passing below transition temperature, the lines of induction $\vec{B}$ are ejected from the sphere

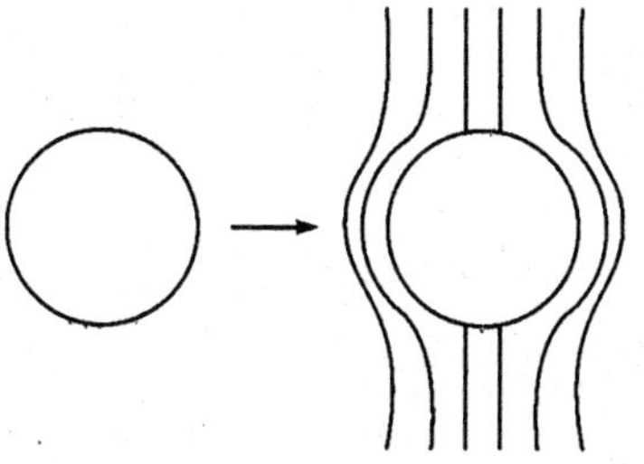

**Figure 8.3(b)** Meissner effect for the specimen initially cooled below transition temperature, then placed in a magnetic field

Meissner effect shows that a superconductor behaves as if inside the specimen $\vec{B} = 0$ at critical temperature, where $\vec{B}$ is the magnetic induction. Hence we can write,

$$\vec{B} = \mu_0 \left(\vec{H} + \vec{M}\right) = 0 \tag{8.2}$$

or, $$\vec{H} = -\vec{M},$$

where $\vec{H}$ is the magnetic intensity and $\vec{M}$ is the magnetization.

$\therefore$ the susceptibility $$\chi = \frac{\vec{M}}{\vec{H}} = -1. \tag{8.3}$$

Therefore, superconductor is true for a perfect diamagnet.

According to Ohm's law, we have

$$\vec{E} = \vec{J}\rho, \tag{8.4}$$

where $\vec{E}$ is the applied electric field vector, $\rho$ the resistivity of the material and $\vec{J}$ the current density. In a superconductor resistivity $\rho$ is zero and $\vec{J}$ has a finite value.

$$\therefore \quad \vec{E} = 0.$$

According to the Maxwell's equation,

$$\vec{\nabla} \times \vec{E} = -\frac{\partial \vec{B}}{\partial t}. \tag{8.5}$$

$$\therefore \quad \vec{\nabla} \times \vec{E} = 0$$

$$\text{or,} \quad \frac{\partial \vec{B}}{\partial t} = 0 \tag{8.6}$$

$$\text{or,} \quad \vec{B} = \text{constant}. \tag{8.7}$$

i.e., the flux through the metal cannot change on cooling through the transition. This result contradicts by Meissner effect. According to Meissner effect the superconductor has perfect diamagnetism and zero resistivity. Thus, condition defining superconducting state is

$$\vec{E} = 0$$

and
$$\vec{B} = 0\,.$$

## 8.3 SPECIFIC HEAT

The specific heat of a normal metal is observed to be of the form:

$$C_n = AT + BT^3 \tag{8.8}$$

The first term on right-hand side of equation (8.8) represents the electronic contribution to the specific heat while the second term on the right-hand side of equation (8.8) represents lattice contribution which is proportional to $T^3$ and is explained by Debye theory. The electronic contribution of the specific heat with temperature for a normal state and superconducting state is plotted in Figure 8.4. The superconducting state shows exponential variation of specific heat with temperature. The presence of magnetic field in the normal state does not significantly modify the contribution to the specific heat by the electrons in the conduction band.

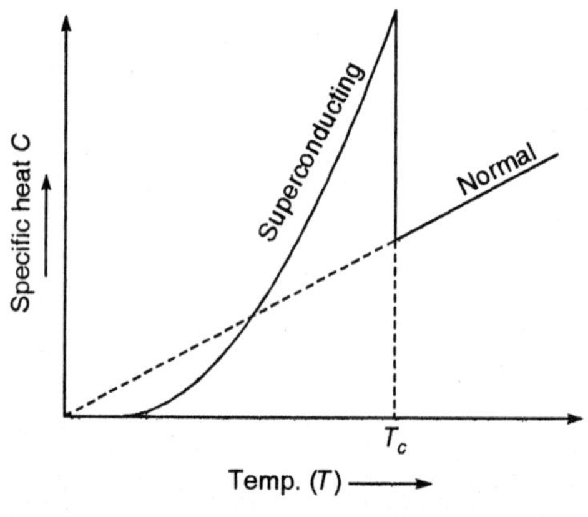


**Figure 8.4** Specific heat versus Temperature dependence of electronic contribution in normal and superconducting state

Thermodynamically it is found that entropy in the normal state always is greater than the entropy in the superconducting state or in other form, the superconducting phase is more ordered than the normal phase. If $S_n$ be the entropy per unit volume of the normal phase and $S_s$ is the entropy per unit volume in superconducting phase, then we can write,

$$S_n - S_s = -\mu_o H_C \frac{dH_e}{dT}, \tag{8.9}$$

where $H_e$ is the critical magnetic field.

The slope $(dH_C/dT)$ versus $T$ curve is always negative (vide Figure 8.5).

$$\therefore \qquad S_n > S_s$$

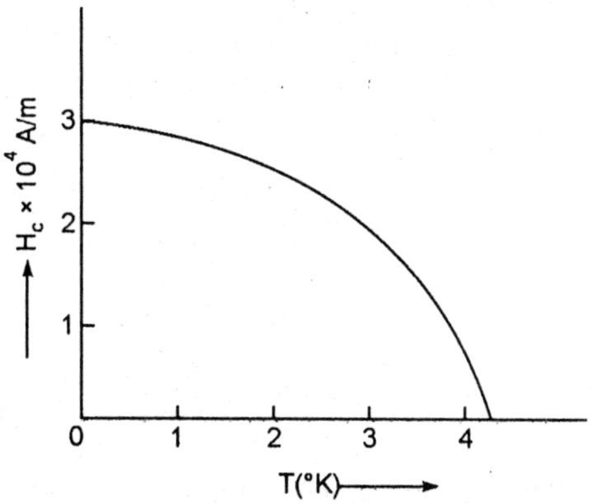


**Figure 8.5** Superconducting variation of $H_c$ with $T$ for $Pb$

Differentiating the equation (8.9) with respect to $T$ and multiplying by $T$, we get

$$T\left(\frac{dS_n}{dT}\right) - T\left(\frac{dS_S}{dT}\right) = -\mu_0 T \frac{d}{dT}\left(H_C \frac{dH_C}{dT}\right)$$

or, $$C_n - C_s = -\mu_0 T \frac{d}{dT}\left(H_C \frac{dH_C}{dT}\right) \tag{8.10}$$

as $$C_n = T\left(\frac{dS_n}{dT}\right) \text{ and } C_S = T\frac{dS_s}{dT},$$

where $C_n$ and $C_s$ are the specific heat per unit volume in normal and superconducting states.

$\therefore$ $$C_s - C_n = \mu_0 T\left(\frac{dH_C}{dT}\right)^2 + \mu_0 T H_C\left(\frac{d^2H_C}{dT^2}\right).$$

Since $$H_C(T_C) = 0 \text{ at } T = T_C,$$

$\therefore$ $$\frac{d^2H_C}{dT^2} = 0.$$

$\therefore$ $$C_s - C_n = \mu_0 T_C\left(\frac{dH_C}{dT}\right)^2_{T=T_C}. \tag{8.11}$$

The equation (8.11) shows an increase of specific heat as the sample is cooled below the critical temperature.

## 8.4 LONDON EQUATION

F. London and H. London developed the macroscopic theory of superconductivity in 1934. The assumption of zero resistivity leads to the acceleration equation as

$$m\frac{d\vec{v}}{dt} = -e\vec{E}, \tag{8.12}$$

where $m$ is the mass of the electron, $\vec{v}$ is the drift velocity at time $t$, $\vec{E}$ the electric field of electron of charge $e$. The superconductor having current density $\vec{J}_s$ is given by

$$J_s = -n_s e\vec{v}, \tag{8.13}$$

where $n_s$ is the number of electrons per unit volume. Differentiating the equation (8.13) with respect to $t$, we have

$$\frac{d\vec{J}_s}{dt} = -n_s e\frac{d\vec{v}}{dt}. \tag{8.14}$$

Substituting the value of $\frac{d\vec{v}}{dt}$ from equation (8.12) in equation (8.14), we get

$$\frac{d\vec{J}_s}{dt} = \frac{n_s e^2}{m}\vec{E} = \frac{\vec{E}}{\lambda}, \tag{8.15}$$

where $\lambda = \frac{m}{n_s e^2}$ is a constant characteristic of superconductor. The equation (8.15) is known as the *first London equation*.

From Maxwell's equation, we know

$$\vec{\nabla}\times\vec{E} = -\frac{\partial\vec{B}}{\partial t} = -\mu_0\frac{\partial\vec{H}}{\partial t}, \tag{8.16}$$

where $\vec{B} = \mu_0\vec{H}$.

Now taking curl of both sides of equation (8.15) and using equation (8.16), we get

$$\vec{\nabla}\times\frac{\partial\vec{J}_s}{\partial t} = \frac{1}{\lambda}\left(\vec{\nabla}\times\vec{E}\right) = -\frac{\mu_0}{\lambda}\frac{\partial\vec{H}}{\partial t}. \tag{8.17}$$

Integrating equation (8.17) with respect to time, we get

$$\vec{\nabla}\times\vec{J}_s = -\frac{\mu_0}{\lambda}\left(\vec{H}-\vec{H}_0\right), \tag{8.18}$$

where $H_0$, the magnetic intensity, is a constant of integration at $t = 0$.

The result admits the particular solution $H = H_0$, but Meissner effect tells us that we cannot have frozen in magnetic fields in the superconductor so that $\vec{H}_0 = 0$ as suggested by $F$ and $H$ Londons; therefore, the equation (8.18) reduces to

$$\vec{\nabla}\times\vec{J}_s = -\frac{\mu_0}{\lambda}\vec{H} \tag{8.19}$$

The equation (8.19) is referred to as the *second London equation*.

## 8.5 LONDON PENETRATION DEPTH

From Maxwell's equation, we know

$$\vec{\nabla}\times\vec{H} = \vec{J}_s + \frac{\partial\vec{D}}{\partial t}, \tag{8.20}$$

where $\vec{D}$ is the electric displacement vector.

Taking curl of equation (8.20), we obtain

$$\vec{\nabla}\times\vec{\nabla}\times\vec{H}=\vec{\nabla}\times\left(\vec{J}_s+\frac{\partial\vec{D}}{\partial t}\right)$$

or, $$\vec{\nabla}\left(\vec{\nabla}.\vec{H}\right)-\nabla^2\vec{H}=\vec{\nabla}\times\vec{J}_s+\vec{\nabla}\times\frac{\partial\vec{D}}{\partial t}$$

Since $$\vec{\nabla}.\vec{H}=0 \text{ and for d.c. case, } \frac{\partial\vec{D}}{\partial t}=0 .$$

$$\therefore \quad -\nabla^2\vec{H}=\vec{\nabla}\times\vec{J}_s . \qquad (8.21)$$

With the help of equation (8.19) the equation (8.21) becomes

$$-\nabla^2\vec{H}=-\frac{\mu_0}{\lambda}\vec{H}$$

or, $$\nabla^2\vec{H}=\frac{\vec{H}}{\lambda_L}, \qquad (8.22)$$

where $$\lambda_L=\frac{\lambda}{\mu_0}.$$

$\lambda_L$ is the dimension of length and is called the *characteristic length* or the *penetration depth* of the field.

Now considering the one-dimensional form of equation (8.22), we have

$$\frac{d^2H}{dx^2}=\frac{H}{\lambda_L} \qquad (8.23)$$

The solution of equation (8.23) is

$$H=H_0\,e^{-\frac{x}{\lambda_L}} .$$

We have discarded the solution having positive exponential power which involves infinite magnetic field. $H_0$ is the value of magnetic field at the surface. $\lambda_L$ is the distance for $H$ to fall from $H_0$ to $H_0/e$. The field falls rapidly with distance $x$ inside the superconductor and is almost zero for a large value of $x$.

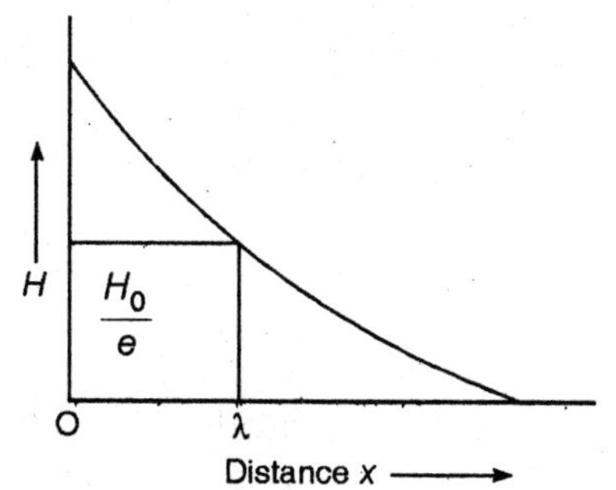


**Figure 8.6** Decay of field with distance

## 8.6 TYPE I AND TYPE II SUPERCONDUCTORS

Superconductors can be divided into two types—Type I and Type II superconductors. This division was done by the magnetic behaviour of an external magnetic field on the superconductor. Type I superconductor is also called *soft superconductor*. These types of superconductors are perfectly diamagnetic and exhibit Meissner effect completely. This type of superconductor exhibits (perfect diamagnetism below a) the critical field $H_C$ (Figure 8.7). When a superconductor is cooled in a magnetic field below the transition temperature, then at the transition temperature for superconductor the magnetic flux originally present are pushed out of the specimen (Figure 8.3). Above the critical field $H_C$ the material behaves as a normal conductor. The value of $H_C$ is very small for type I superconductor as they do not have technical application. A few examples of Type I superconductor are Sn, Pb, Zn, Cd.

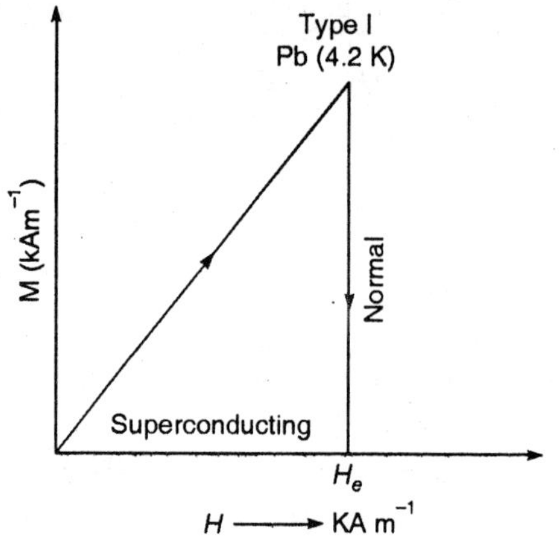


**Figure 8.7** Magnetization vs. magnetic field of pure Pb, critical field at 4.2 °K

In Type II superconductors the superconductivity is partially destroyed for $H_{C1} \leq H \leq H_{C2}$ as in Figure 8.8. The region is called the intermediate state since it is really a mixture of normal and superconducting states. A few examples of Type II superconductor are $Nb_3Ge$, $Nb_3Sn$. In general, the intermediate state is unstable in Type I superconductor but stable for Type II or hard superconductor.

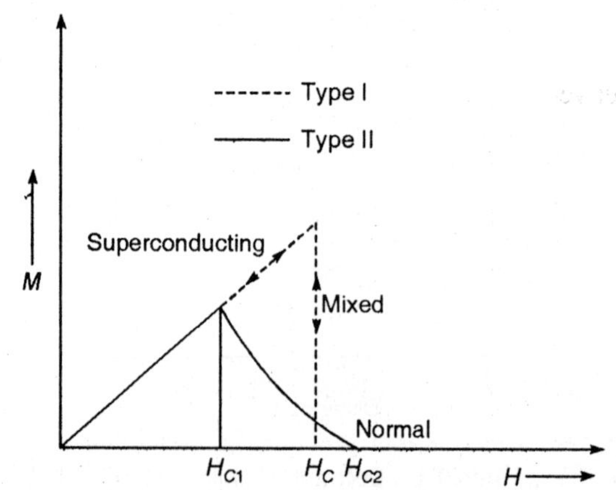


**Figure 8.8** The plot of magnetization (M) versus magnetic field for Type I and Type II superconductors

## 8.7 BCS THEORY OF SUPERCONDUCTIVITY

The theory of superconductivity was sucessfully explained by Bardeen, Cooper and Schrieffer in 1957 and in short known as BCS theory after the intials of the proposers' names. According to BCS theory after the intial interaction responsible for superconductivity appears to be that of a pair of electrons by means of an interchange of virtual phonons. We know that two electrons in free space repell each other by the coulomb force but in a metal when an electron moves through the lattice it distorts the lattice, this distortion gives rise to a phonon. A second electron at some distance away is in turn affected when it is reached by the propagating fluctuation in the lattice change distribution. The interaction between the lattice and phonon is the form of constant emission and absorption of virtual phonons by electrons.

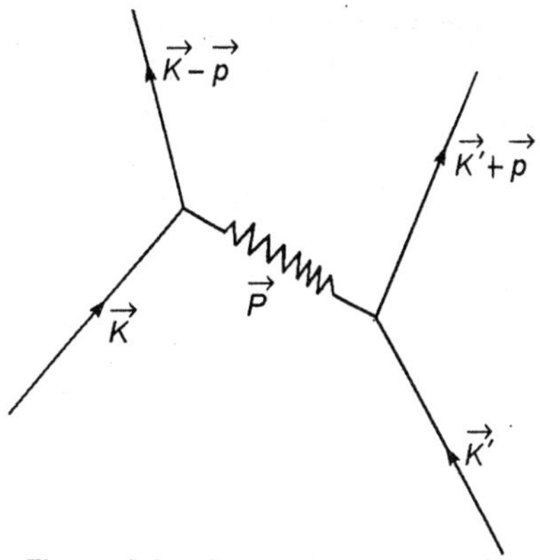


**Figure 8.9** Electron-phonon-electron interaction

The basic assumption of the BCS theory is that when superconductivity occurs there is such an attractive interaction between two electrons by means of virtual phonon exchange that dominates the usual repulsive Coulomb interactions.

In Figure 8.9 we have seen electron-electron interaction by virtual phonon exchange. In Figure 8.9, it is shown that an electron in the state $\vec{K}$ emits a virtual phonon $\vec{p}$ and is scattered in the state $\vec{K}-\vec{p}$. The electron in state $\vec{K}'$ absorbs this virtual phonon and is scattered into $\vec{K}' + \vec{p}$. The process being a virtual one energy need not be conserved. Such a pair of electrons is called *Cooper pair* as it was first discovered by L.N. Cooper. Cooper in 1965 showed that such a pair of electrons can be regarded as single particle and forms a bound state. The energy of such a pair of electrons in the bound state is less than the energy of two unbound electrons. Cooper pair obeys Bose-Einstein statistics and can be in the lowest energy state. It was shown by BCS that energy of the system is the lowest when the total momentum of each pair is same and is zero. In this state the electron pairs condense and the flow of Cooper pairs constitutes the supercurrent.

## 8.8 TUNNELLING

Giaever in 1960 measured the energy gap by considering a situation in which one side of the tunnel junction exists between a normal metal and a superconductor. Actually he was observing

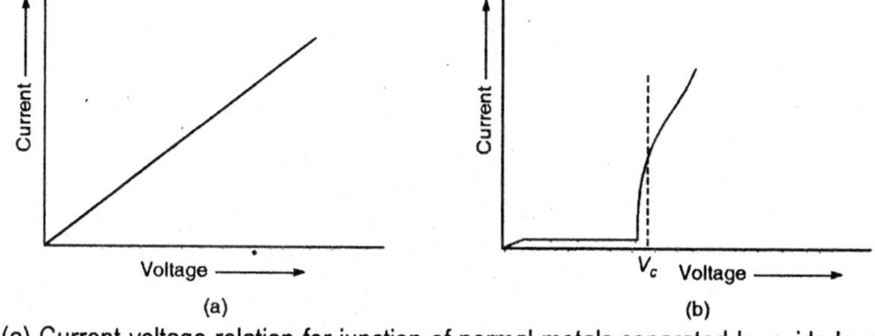


**Figure 8.10** (a) Current-voltage relation for junction of normal metals separated by oxide layer, (b) Current-voltage relation with one metal normal and other metal superconducting

the tunnelling of electrons between superconducting film and a normal metal separated by a thin insulating barrier. If the barrier is sufficiently thin there is a finite probability that an electron which impinges on the barrier will pass from metal to the superconductor. This is called *tunnelling*. The current-voltage relation with one metal normal and other metal superconducting is shown in Figure 8.10. The energy diagram displaying the density of states is shown in Figure 8.11.

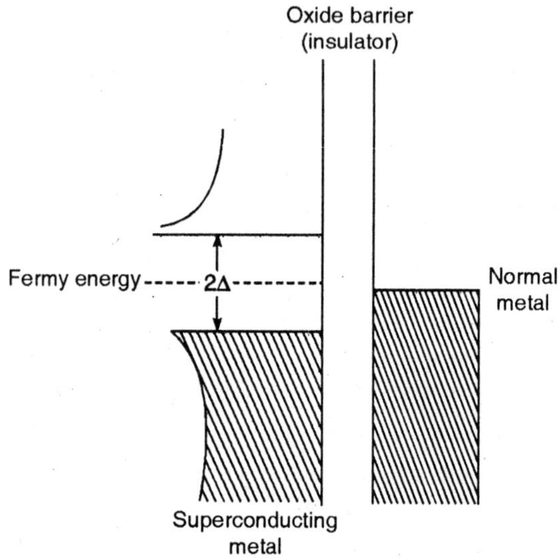


**Figure 8.11** Energy diagram displaying density states one in normal metal and the other in the superconducting state

At absolute zero no current can flow until the applied voltage is $V = E_g/2e = \Delta/e$, which is half the energy gap. At temperature different from zero there is a small current flow even at low voltage because of the electrons in the superconductor which are thermally excited across the energy gap. The superconducting films are then compared to the penetration depth and because of their size their critical field is very high. Single electron tunnelling in superconductor has been highly successful and the results are in good agreement with the BCS theory.

## 8.9 JOSEPHSON EFFECTS

Josephson observed that tunnelling of Cooper pairs could take place between two superconductors sandwiched by a very thin insulator and this effect takes place even without a potential difference between superconductors (Figure 8.12). There are two types of Josephson effect.

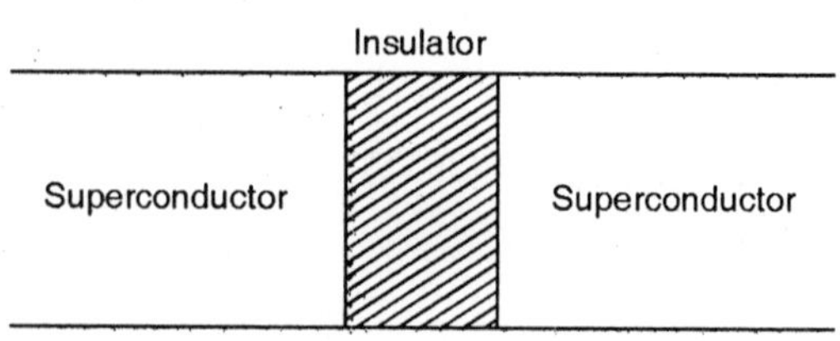


**Figure 8.12** Josephson junction for two superconductors separated by a thin insulator

(i) The d.c. Jospehson effect and

(ii) the a.c. Jospehson effect.

(i) According to the d.c. Josphson effect we get current with no applied voltage but no current with an applied voltage.

(ii) The a.c. Josphson effect occurs when both a.c. and d.c. voltages of high frequency are applied across the junction. The RF current oscillation of frequency $f = qV/h = 2eV/h$, where Cooper pair of charge $q = 2e$. The a.c. Josephson effect can be used to measure the fundamental constant $e/h$ very accurately.

*Superconducting quantum interference device* (SQUID) occurs in magnetic field effect. A SQUID consisting of a ring of superconductor having two Josephson junction with two side-arms $a$ and $b$ acts as an entrance and exit for supercurrent respectively. The magnetic field is applied to the central region.

The insulating layers 1 and 2 may in general have different thickness and let the current flow through these paths $J_1$ and $J_2$ respectively and total current $J_T = J_1 + J_2$. Both $J_1$ and $J_2$ vary periodically with magnetic field. Since the current is sensitive to very small changes in magnetic field, the SQUID can be used as a galvanometer. It is used to measure magnetic field also.

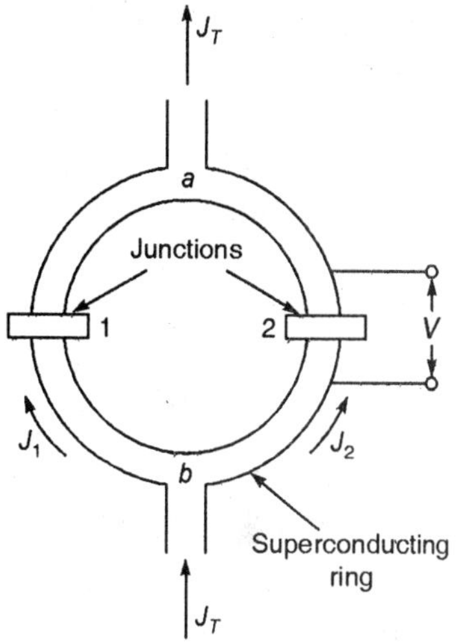


**Figure 8.13** Superconducting quantum interference device (SQUID) formed by two Josephson junctions 1 and 2

## 8.10 APPLICATIONS OF SUPERCONDUCTOR

Superconductivity has been used in many fields of engineering and radioelectronics. With the help of Josephson effect we can measure very accurately the constant $e/h$. Superconducting quantum interference device (SQUID) can be used as a very sensitive galvanometer and it is also used to measure magnetic field. The superconducting magnets have been utilised in NMR spectrometers and NMR imaging used in medical diagnosis. It can also be used in highly powerful strong superconductor electromagnets. The new branch of electronics based on superconductivity is known as *cryoelectronics*. With this branch the miniature receivers capable of detecting extremely weak radio signal are created. The manufacture of large scale integrated (LSI) circuit and very large scale integrated (VLSI) circuit for the development of supercomputer is possible. The cryogenic devices belong memories and logic unit of the computer engineering.

The resistance of a metal in normal and superconducting states is infinite. Superconductor can be used as a switching device. This type of switch is known as *cryotron switch*. Low temperature superconductor has limited use because of the requirement of liquid helium temperature which is a great economic disadvantage.

## SOLVED PROBLEMS

**1.** Lead in superconducting state has critical temperature of 6 °K at zero magnetic field and a critical field of 0.064 MAm$^{-1}$ at 0 °K. Determine the critical field at 4 °K.

**Soln.** We know from the equation (8.1) the critical field can be expressed as

$$H_C = H_0\left[1-\left(\frac{T}{T_C}\right)^2\right].$$

Here $H_0$ = 0.064 MAm$^{-1}$, $T$ = 4 °K and $T_C$ = 6 °K

$$\therefore \qquad H_C = 0.064\left[1-\left(\frac{4}{6}\right)^2\right] \text{M Am}^{-1}$$

$$= \mathbf{0.0355\ MAm^{-1}}.$$

**2.** The critical fields at 6 °K and 8 °K for NbTi alloy are 7.616 MAm$^{-1}$ and 4.284 MAm$^{-1}$ respectively. Determine the transition temperature and critical field at 0 °K.

**Soln.** We know from the equation,

$$H_C = H_0\left[1-\left(\frac{T}{T_C}\right)^2\right].$$

From the problem, we get

$$7.616 = H_0\left[1-\left(\frac{6}{T_0}\right)^2\right] \qquad (1)$$

$$4.284 = H_0\left[1-\left(\frac{8}{T_0}\right)^2\right]. \qquad (2)$$

Dividing equation (1) by equation (2), we get

$$\frac{7.616}{4.284} = \frac{1-(6/T_0)^2}{1-(8/T_0)^2}. \qquad (3)$$

Solving the equation (3), we get

$$T_0 = \mathbf{10\ °K}.$$

Now substituting the value of $T_0$ in equation (1), we get

$$7.616 = H_0\left[1-\left(\frac{6}{10}\right)^2\right].$$

$$\therefore \qquad H_0 = \mathbf{11.9\ MAm^{-1}}.$$

**3.** The critical temperature $T_C$ for mercury with isotopic mass 199.5 amu is 4.185 °K. Calculate its $T_C$ when its mass changes to 203.4 amu.

**Soln.** The critical temperature of superconductor is related to its isotopic mass as

$$T_C \propto \frac{1}{\sqrt{M}}. \qquad (1)$$

If $T_{C1}$ is the critical temperature of mass $M_1$ and $T_{C2}$ is the critical temperature of mass $M_2$, then from equation (1) we get

$$\frac{T_{C1}}{T_{C2}} = \sqrt{\frac{M_2}{M_1}}.$$

$$\therefore \qquad T_{C1} = T_{C2}\sqrt{\frac{M_2}{M_1}}.$$

From the problem, we get

$$T_{C1} = 4.185\ °\text{K},\ M_1 = 199.5 \text{ amu and } M_2 = 203.4 \text{ amu}.$$

$$\therefore \qquad 4.185 = T_{C2}\sqrt{\frac{203.4}{199.5}}$$

$$\therefore \qquad T_{C2} = \mathbf{4.145\ °K}.$$

## REVIEW QUESTIONS

1. What are superconductors? Give a few examples.
2. What do you mean by transition temperature and critical field for superconductor?
3. What is Meissner effect? Calculate the magnetic susceptibility of a superconductor.
4. What is the effect of magnetic field on superconductor?
5. How does the specific heat of a superconductor depend on temperature?
6. Deduce London equation.
7. Establish the relation: $H = H_0\, e^{-x/\lambda_L}$, where $\lambda_L$ is the penetration depth.
8. What do you mean by Type I and Type II Superconductors?
9. Write a note on BCS theory.
10. What is Cooper pair?
11. Write a note on Josephson effect.
12. What is a SQUID?
13. Write a note on a few applications of superconductivity.
14. How do the electrical, magnetic and thermodynamic properties of superconductors differ from those of normal conductors?

## PROBLEMS

1. For lead the critical field at 0 °K is $6.39 \times 10^4$ A/m and the critical temperature for zero magnetic field is 7.18 °K. Find the critical field for lead at 4 °K. [**Ans.** $4.41 \times 10^4$ A/m]
2. The critcal fields of a superconductor at 6 °K and 8 °K are $7.616 \times 10^6$ A/m $4.284 \times 10^6$ A/m respectively. Find the critical temperature for zero magnetic intensity and critical field at 0 °K. [**Ans.** 10 K, $11.9 \times 10^6$ A/m]
3. Mercury having an average atomic mass of 200.59 amu has a critical temperature of 4.153 °K. Calculate critcal transition temperature of the isotopic $_{80}Hg^{204}$. [**Ans.** 4.118 K]

# 9

# *Dielectric Properties of Solids*

## 9.0 INTRODUCTION

Matter comes in many varieties, viz., solid, liquid and gases and all these substances do not respond in the same way to electrostatic field. The medium may be classified as two large classes (1) conductor and (2) insulator. The medium in which the static electrc field can exist is called *insulator* or *dielectric.* A dielectric is an electrical insulation that can be polarized by an applied electric field. A substance whose basic electrical property is the ability to be polarized and in which an electrostatic field can exist is called *dielectric material.* If a dielectric is composed of weakly bound molecules, those molecules not only become polarized, but also reorient so that their symmetry axes align to the field. Dielectric is used to describe material with a high polarizability.

The study of dielectric is concerned with storage and dissipation of electric and magnetic energy in material. It is important to explain various phenomena in electronics, optics and solid state physics. In the dielectric all charges are attached to specific atoms or molecules and they are on a tight bound together and they can move within the atoms or molecules. The electric field can distort the charge distribution of a dielectric atom or molecules. There are practically no free electrons in dielectrics. A few examples of dielectrics are sulphur, porcelain, mica, glass, ebonite, paraffin, etc. The dielectric properties are utilised in fabricating various devices such as capacitor, microphone, etc.

## 9.1 ELECTRIC POLARIZATION

When a rectangular block of dielectric is placed in an external electric field $\vec{E}_0$ under the action of the electric field the positive charge on the molecule of the dielectric is slightly displaced in the direction of the field while the negative charge is displaced in the opposite direction as in Figures 9.1(a) and 9.1(b). The separation of charge is proportional to the intensity of the electric field. Two equal and opposite charges separated by a distance constitute an electric dipole. The product of the either charge and the distance separating them is called *electric dipole moment*. The atom has a tiny dipole moment $\vec{p}$ which points in the same direction as $\vec{E}_0$. This induces dipole moment proportional to the field $\vec{E}_0$, so

$$\vec{p} = \alpha \vec{E}_0 . \tag{9.1}$$

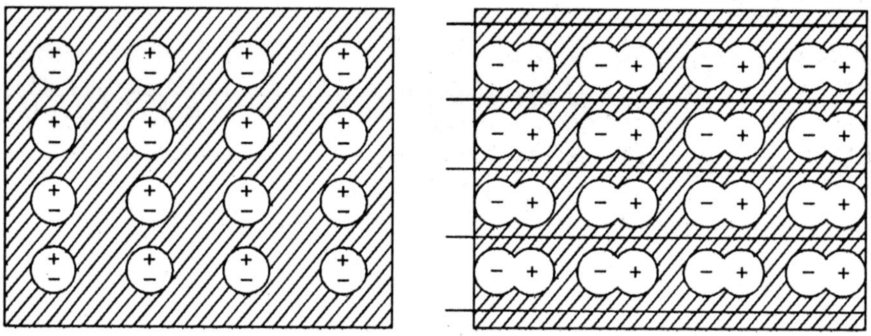

(a) (b)

**Figure 9.1**

The constant of proportionality $\alpha$ is called *atomic polarizability*.
From equation (9.1), we get

$$\alpha = \frac{\vec{p}}{\vec{E}_0} .$$

The atomic polarizability $\alpha$ is defined as the atomic dipole moment induced in an atom by an electric field of unit strength.

$$\text{Unit of } \alpha = \frac{\text{Cm}}{\text{Vm}^{-1}} = \text{Fm}^2 .$$

The result of a large number of dipole each having same dipole moment is in the direction of the field. The dipole moment per unit volume of the material is said to be polarization $\vec{P}$ of the medium. The effect of polarization is to reduce the external field $\vec{E}_0$. The resultant field

$$\vec{E} = \vec{E}_0 + \vec{E}_p , \tag{9.2}$$

where $\vec{E}_p$ is the polarization field and it is in the opposite direction to applied field and $\vec{E} < \vec{E}_0$.

For ordinary electric field, polarization $\vec{P}$ is proportional to $\vec{E}$

or, $\quad \vec{P} \propto \vec{E}$

$\therefore \quad \vec{P} = \text{const.} \times \vec{E} .$

In the simplest case of a homogeneous field with volume of dielectric $v$,

$$\vec{P} = \frac{\Sigma \vec{p}}{v}.$$

In S.I. unit constant = $\varepsilon_0 x_e$.

$$\therefore \qquad \vec{P} = \varepsilon_o x_e \vec{E}, \tag{9.3}$$

where $\varepsilon_o$ is the electric permittivity of free space and $x_e$ is the electric susceptibility, the product $\varepsilon_0 x_0$ is known as *absolute dielectric susceptibility.*

## 9.2 LOCAL FIELD

In solids the distance between the molecules is small, so a dipole sees not only the external field but also the fields created by the dipoles of other particles. The actual electric field acting on atoms or molecules is called *local field* or *internal field.* This field is responsible for polarization of each atom or molecule. For an atomic site with cubic symmetry the local field is given by *Lorentz relation* as

$$\vec{E}_{loc} = \vec{E}_o + \vec{E}_P + \frac{\vec{P}}{3\varepsilon_o} = \vec{E} + \frac{\vec{P}}{3\varepsilon_o} \tag{9.4}$$

as $\vec{E} = \vec{E}_o + \vec{E}_p$ in equation (9.2).

The expression (9.4) gets modified with the shape of the specimen. $\vec{E}_{10c}$ contains a term due to polarization.

## 9.3 CLAUSIUS-MOSSOTTI RELATION

When an electric field $\vec{E}$ is applied to a material, relative displacement of positive and negative charges takes place. This displacement induces an electric dipole moment which is expressed in terms of polarization $\vec{P}$. The electric displacement vector for an isotropic or cubic medium is defined as

$$\vec{D} = \varepsilon_o \varepsilon_x \vec{E} = \varepsilon_o \vec{E} + \vec{P}. \tag{9.5}$$

where $\varepsilon_x$ is called the *dielectric constant* or *relative permittivity* of the medium. From equation (9.5), we get

$$\varepsilon_x = \frac{\varepsilon_o \vec{E} + \vec{P}}{\varepsilon_o \vec{E}} = 1 + \frac{\vec{P}}{\varepsilon_o \vec{E}}. \tag{9.6}$$

From equation (9.3), we can write, $x_e = \dfrac{\vec{P}}{\varepsilon_o \vec{E}}$.

Therefore, the equation (9.6) transforms to

$$\varepsilon_x = 1 + x_e. \tag{9.7}$$

The atomic polarizability $\alpha$ of an atom is given from equation (9.1)

$$\alpha = \frac{\vec{p}}{\vec{E}_{loc}}. \tag{9.8}$$

where $\vec{p}$ is the dipole moment of the atom and $\vec{E}_{loc}$ is the local electric field at the atom. Here in equation (9.1) $\vec{E}_0$ is replaced by $\vec{E}_{loc}$, which is actual electric field acting on atoms or molecules. $\alpha$ is defined as an atomic or ionic property whereas $x_e$ and $\varepsilon_r$ will depend on the manner in which the atoms are assembled in the crystal. If all the atoms have the same polarizability and there are $N$ numbers of atoms per unit volume, then *polarization* can be expressed with the help of equation (9.8) as

$$\vec{P} = N\vec{p} = N\,\alpha\,\vec{E}_{loc}\,. \tag{9.9}$$

Taking magnitude only, the polarization is the dipole moment per unit volume

$$P = \sum_i N_i\,\alpha_i\,E_{loc}(i)\,, \tag{9.10}$$

where $N_i$ is the number of atoms per unit volume of polarizability $\alpha_i$ and $E_{loc}(i)$ is the local field at atoms of type $i$.

In the case of cubic or isotropic structures using equations (9.4) and (9.9), we get

$$P = N\,\alpha\left(E + \frac{P}{3\varepsilon_o}\right)$$

or,

$$P\left(1 - \frac{N\alpha}{3\varepsilon_o}\right) = N\,\alpha\,E$$

or,

$$\frac{P}{E} = \frac{N\alpha}{1 - \dfrac{N\alpha}{3\varepsilon_o}}\,. \tag{9.11}$$

Dividing both sides by $\varepsilon_o$, the equation (9.11) becomes

$$\frac{P}{\varepsilon_o E} = \frac{\dfrac{N\alpha}{\varepsilon_o}}{1 - \dfrac{N\alpha}{3\varepsilon_o}}\,. \tag{9.12}$$

Using equation (9.3), we can write the equation (9.12) as

$$x_e = \frac{\dfrac{N\alpha}{\varepsilon_o}}{1 - \dfrac{N\alpha}{3\varepsilon_o}}\,. \tag{9.13}$$

From equation (9.7) the equation (9.13) is

$$\varepsilon_r = 1 + \frac{\dfrac{N\alpha}{\varepsilon_o}}{1 - \dfrac{N\alpha}{3\varepsilon_o}}$$

$$= \frac{1 + \dfrac{2N\alpha}{3\varepsilon_o}}{1 - \dfrac{N\alpha}{3\varepsilon_o}}.$$

$$\therefore \qquad \varepsilon_r - 1 = \frac{\dfrac{N\alpha}{\varepsilon_o}}{1 - \dfrac{N\alpha}{3\varepsilon_o}}$$

and

$$\varepsilon_r + 2 = \frac{3}{1 - \dfrac{N\alpha}{3\varepsilon_o}}.$$

Therefore,

$$\frac{\varepsilon_r - 1}{\varepsilon_r + 2} = \frac{N\alpha}{3\varepsilon_o}. \tag{9.14}$$

This is Claussius-Mossotti relation. It gives a relation between polarizability and dielectric constant of a solid.

For a non-magnetic material the refractive index

$$n = \sqrt{\varepsilon_r} \quad \text{or,} \quad n^2 = \varepsilon_r.$$

Therefore, Claussius-Mossotti relation transforms as

$$\frac{n^2 - 1}{n^2 + 2} = \frac{N\alpha}{3\varepsilon_o}. \tag{9.15}$$

In more general way, the equation (9.14) can be expressed as

$$\frac{\varepsilon_r - 1}{\varepsilon_r + 1} = \frac{1}{3\varepsilon_0} \sum_{\varepsilon} N_i \alpha_i .$$

## 9.4 TYPES OF POLARIZABILITY

The net polarizability of a dielectric has the following contributions:

1. **Electronic polarizability $\alpha_{\text{electronic}}$.** This is due to the displacement of electrons in an atom relative to the nucleus by an external electric field as in Figure 9.2.

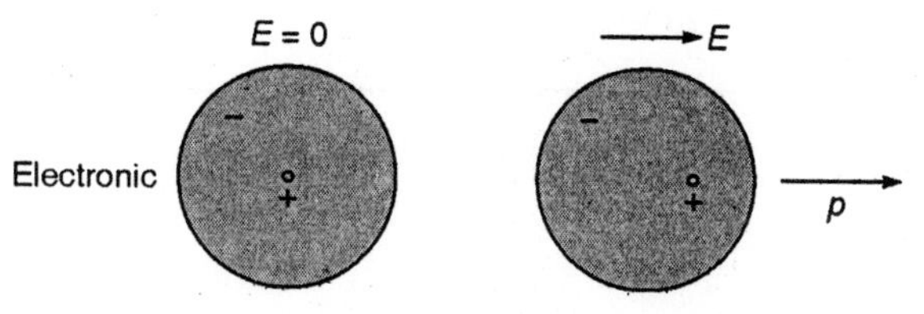


**Figure 9.2** Depicts electronic polarizability

2. **Ionic polarizability $\alpha_{ionic}$.** This is due to the displacement of ions with respect to each other by an external electronic field as in Figure 9.3.

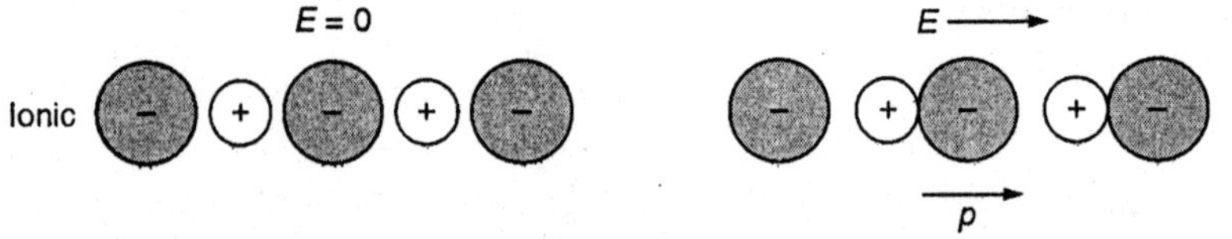


**Figure 9.3** Depicts ionic polarizability

3. **Dipolar or orientational polarizability $\alpha_{dipolar}$.** This is due to the molecules with permanent electric dipole moment. These molecules align themselves in the presence of external electric field as in Figure 9.4.

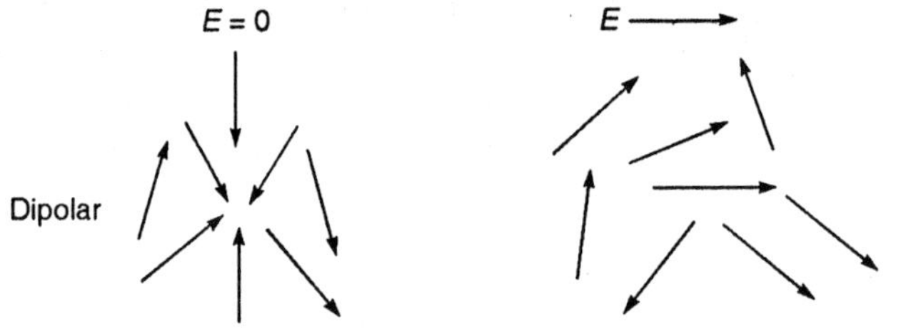


**Figure 9.4** Dipicts dipolar or orientational polarizability

Therefore, total polarizability,

$$\alpha = \alpha_{dipolar} + \alpha_{ionic} + \alpha_{electronic}.$$

Total polarizability $\alpha$ with frequency is plotted in Figure 9.5.

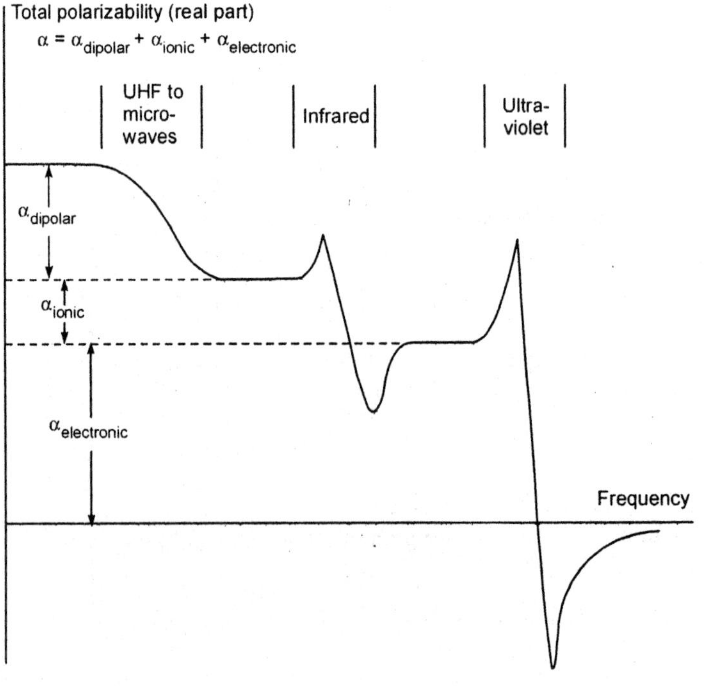


**Figure 9.5** Dipicts total polarizability with frequency

The extent to which a particular polarizability contributes depends on the nature of the dielectric and the frequency of the electric field.

**1. Electronic Polarizability**

Electronic polarizability arises due to the displacement of electrons in an atom relative to the nucleus by an external electric field.

*Classical theory of electronic polarizability*: An electron bound harmonically to an atom will show resonance absorption at a frequency $\omega_o = \left(\frac{\beta}{m}\right)^{1/2}$, where $\beta$ is the force constant. The displacement $x$ of the electron produced by the application of a field $E_{loc}$ is given by

$$eE_{loc} = \beta x = m\omega_o^2 x\,. \tag{9.16}$$

The static electronic polarizability,

$$\alpha_{\text{electronic}} = \frac{p}{E_{loc}} = \frac{ex}{E_{loc}} = \frac{e^2}{m\omega_o^2}\,. \tag{9.17}$$

The electronic polarizability will depend on frequency as given below:

If $\omega$ is frequency of local field and the field at any time $t$ is given by $E_{loc}$ sin $\omega t$, the equation of motion is

$$m\frac{d^2x}{dt^2} + m\omega_o^2 x = -eE_{loc}\sin\omega t\,. \tag{9.18}$$

Now substituting $x = x_0 \sin \omega t$ in equation (9.18), we get

$$m(-\omega^2 + \omega_o^2)x = -eE_{loc}\,.$$

The dipole moment has the magnitude

$$p_o = -ex_o = \frac{e^2 E_{loc}}{m(\omega_o^2 - \omega^2)}\,. \tag{9.19}$$

Therefore, the electronic polarizability is

$$\alpha_{\text{electronic}} = \frac{p_0}{E_{loc}} = \frac{e^2/m}{(\omega_o^2 - \omega^2)}\,. \tag{9.20}$$

In visible region the frequency dependence is not usually important in most dielectric. In quantum theory, the expression corresponding to equation (9.20) is

$$\alpha_{\text{electronic}} = \frac{e^2}{m}\sum_j \frac{f_{ij}}{\omega_o^2 - \omega^2}\,. \tag{9.21}$$

If damping mechanism is included, the general expression for polarizability is

$$\alpha_{\text{electronic}} = \frac{e^2/m}{(\omega_o^2 - \omega^2) + i\omega\beta/m} \tag{9.22}$$

$$= \alpha'_{\text{electronic}} - i\alpha''_{\text{electronic}}\,,$$

where
$$\alpha'_{\text{electronic}} = \frac{e^2(\omega_o^2 - \omega^2)/m}{(\omega_o^2 - \omega^2)^2 + \left(\frac{\omega\beta}{m}\right)^2} \tag{9.23}$$

and
$$\alpha''_{\text{electronic}} = \frac{e^2\omega\beta}{(\omega_o^2 - \omega^2)^2 + \left(\frac{\omega\beta}{m}\right)^2} \tag{9.24}$$

### 2. Ionic Polarizability

This is due to the displacement of ions with respect to each other by an external electric field. In ionic crystal such as KCl, KBr, NaCl the relative displacement of positive ions with respect to negative ions in the presence of electric field gives rise to polarization. Ionic polarization involves the motion of ions which are heavier than electrons so that resonance frequency is in the infrared region. So the ionic contribution is important at low frequencies.

Assuming the forces near equilibrium as simple harmonic the displacement $\Delta x$ by an applied electric field $E$ is given by

$$\beta \Delta x = eE,$$

where $\beta$ is the force constant.

Therefore, ionic polarizability

$$\alpha_{\text{ionic}} = \frac{P}{E} = \frac{e\Delta x}{E} = \frac{e^2}{\beta}. \tag{9.25}$$

### 3. Dipolar or Orientational Polarizability

The polarizability arising from the orientation in applied electric field possesses a permanent dipole moment. The dipolar polarizability is the property of dipolar substance.

When the electric field $\vec{E}$ is present on a permanent dipole in opposite to thermal agitations the dipole orients themselves along the field and produces orientational or dipolar polarizability. We consider the effect of the thermal motion on molecules which are free to rotate. The potential energy $U$ of a molecule of permanent moment $\vec{p}$ in the field $\vec{E}$ is

$$U = -\vec{p}.\vec{E} = -pE\cos\theta, \tag{9.26}$$

where $\theta$ is the angle between moment and the direction of the field.

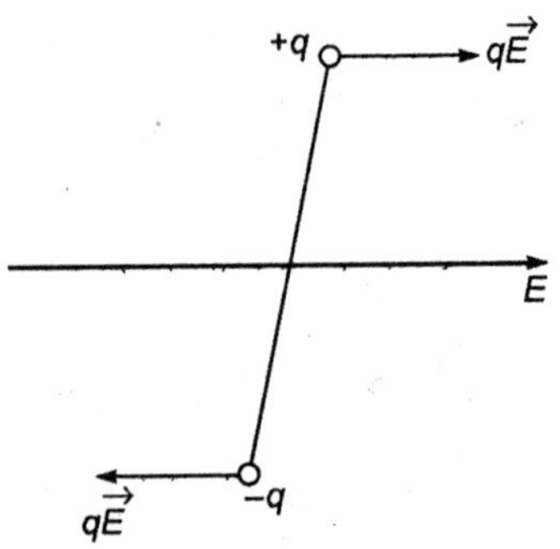


**Figure 9.6** A dipole $\vec{p}$ moment placed in an electric field $\vec{E}$

The probability for a dipole having orientation between $\theta$ and $\theta + d\theta$ is given by Boltzmann distribution law as

$$e^{-U/KT}\,d\Omega$$
$$= 2\pi \sin\theta\, d\theta\, e^{-U/KT}$$
$$= 2\pi \sin\theta\, d\theta\, e^{PE\cos\theta/KT},$$

where the solid angle $d\Omega = 2\pi \sin\theta\, d\theta$.

Contribution made by all dipoles between $\theta$ and $\theta + d\theta$ is

$$= P\cos\theta\, 2\pi \sin\theta\, d\theta\, e^{PE\cos\theta/KT} \quad (9.27)$$

Average contribution per dipole is given by

$$\vec{p} = \frac{\int_0^{\pi} P\cos\theta\, 2\pi\sin\theta\, d\theta\, e^{PE\cos\theta/KT}}{\int_0^{\pi} 2\pi\sin\theta\, d\theta\, e^{PE\cos\theta/KT}} \quad (9.28)$$

The integration is carried out over all solid angles.

To solve the equation (9.28) we put

$$x = \frac{PE}{KT} \text{ and } y = \cos\theta.$$

$$\therefore \quad dy = -\sin\theta\, d\theta.$$

Therefore, the equation (9.28) becomes

$$\vec{p} = \frac{p\int_{+1}^{-1} ye^{xy}\,dy}{\int_{+1}^{-1} e^{xy}\,dy}$$

or, $$\frac{\vec{p}}{p} = \frac{e^x + e^{-x}}{e^x - e^{-x}} - \frac{1}{x} = \coth x - \frac{1}{x} = L(x).$$

$L(x)$ is called *Langevin function*. The variation of $L(x)$ with $x$ is shown in Figure 9.7. Expanding

$\coth x$, $\coth x = \frac{1}{x} + \frac{x}{3} - \frac{x^3}{45} \cdots$

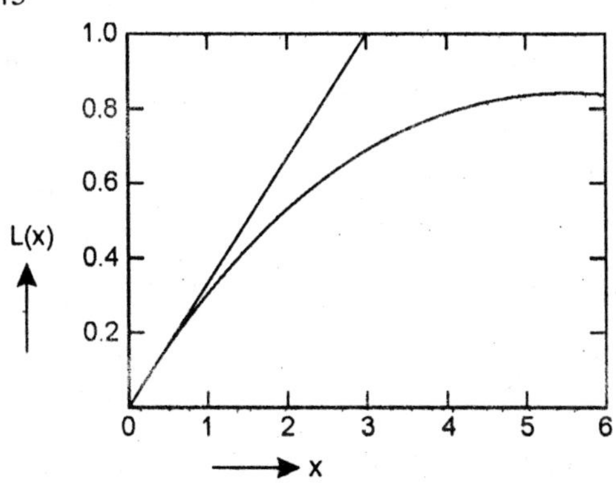


**Figure 9.7** Plot of Langevin function $L(x)$ versus $x$. The initial slope is shown by straight line

For $x << 1$, i.e., $PE << KT$, we have

$$L(x) = \frac{1}{x} + \frac{x}{3} - \frac{1}{x}$$

$$= \frac{x}{3}.$$

Then polarization

$$\frac{\vec{p}}{p} = \frac{x}{3} = \frac{PE}{3KT}$$

or,

$$\vec{p} = \frac{P^2 E}{3KT}.$$

$\therefore$ dipolar polarizability $\frac{\vec{p}}{E} = \frac{p^2}{3KT}$.

## 9.5 DIELECTRIC RELAXATION AND LOSS

Dielectric relaxation is the momentary delay or lag in dielectric constant of a material. This is usually caused by the delay in molecular polarization with respect to a changing electric field in a dielectric medium. The relaxation of a disturbed system to an equilibrium configuration, the relaxation frequency is the reciprocal of the relaxation time. Dielectric relaxation in changing electric field could be considered analogous to hysteresis in changing magnetic fields. Relaxation in general is a delay in the response of a linear system and, therefore, the dielectric relaxation is measured.

## 9.6 FERROELECTRIC CRYSTAL

A ferroelectric crystal exhibits an electric dipole moment even in the absence of electric field. In the ferroelectric state the centre of positive charge of the crystal does not coincide with the centre of negative charge. So in the ferroelectric crystal the phenomenon of spontaneous polarization takes place in the absence of the electric field, i.e., at $\vec{E} = 0$.

Ferroelectricity disappears above a certain temperature called *transition temperature* or the *Curie point* $T_c$, above the transition temperature the crystal is in the paraelectric state. The variation of polarization versus electric field is not linear but forms a loop called *hysterisis loop*. A few known ferroelectric crystals are Rochelle salt NaK $(C_4H_4O_6)4H_2O$, Barium titanate $BaTiO_3$. Thermal motion tends to destroy the ferroelectric order.

## 9.7 PIEZOELECTRICITY

Piezo is a Greek word *piezein* meaning 'pressure'. Therefore, piezoelectricity means pressure electricity, i.e., by application of external stress in certain crystals an electric polarization is produced. Crystals in which electric polarization occurs is called *piezoelectric crystals* and the phenomenon is known as *Piezoelectricity*. The application of an electric field produces mechanical strain in the crystal. This is the inverse effect of piezoelectricity. Some common piezoelectric crystals are quartz, tourmaline and Rochelle salt.

## SOLVED PROBLEMS

1. The relative permittivity of germanium is 16. The edge length of the conventional cubic cell for germanium lattice is $5.65 \times 10^{-10}$ m. Calculate the electronic polarizability of germanium atoms.

**Soln.** From Claussius-Mossotti relation, we know,

$$\frac{\varepsilon_r - 1}{\varepsilon_r + 2} = \frac{N\alpha_e}{3\varepsilon_o}.$$

There will be 8 germanium atoms in the conventional cubic cell of volume $(5.65 \times 10^{-10})^3$ m$^3$. The number of atoms per unit volume $N$ is given by

$$N = \frac{8}{(5.65 \times 10^{-10})^3} = 4.4355 \times 10^{28}.$$

Since $\varepsilon_r = 16$, we get from Claussius-Mossotti relation,

$$\frac{16-1}{16+2} = \frac{4.4355 \times 10^{28}\,\alpha_e}{3 \times 8.854 \times 10^{-12}}$$

or, $\boldsymbol{\alpha_\varepsilon = 4.99 \times 10^{-40}}$ **Fm$^2$.**

2. The crystal of sodium chloride has static dielectric constant of 5.6 and optical index of refraction 1.5. Calculate the percentage contribution of ionic polarizability.

**Soln.** From Claussius-Mossotti relation, we get

$$\frac{\varepsilon_r - 1}{\varepsilon_r + 2} = \frac{N\alpha}{3\varepsilon_o}. \qquad (9.29)$$

If $\alpha_{\text{electronic}}$ is the electronic polarizability and $\alpha_{\text{ionic}}$ is the ionic polarizability, then

$$\alpha = \alpha_{\text{electronic}} + \alpha_{\text{ionic}}. \qquad (9.30)$$

From relation (9.29) and (9.30) we can write,

$$\frac{\varepsilon_r - 1}{\varepsilon_r + 2} = \frac{N(\alpha_{\text{electronic}} + \alpha_{\text{ionic}})}{3\varepsilon_o}. \qquad (9.31)$$

At optical frequencies $\alpha_{\text{ionic}}$ is very small, so $\varepsilon_x = n^2$, where $n$ is the optical refractive index. So the equation (9.31) becomes

$$\frac{n^2 - 1}{n^2 + 2} \times \frac{\varepsilon_r + 2}{\varepsilon_r - 1} = \frac{\alpha_{\text{electronic}}}{\alpha_{\text{electronic}} + \alpha_{\text{ionic}}}$$

or, $$1 - \frac{\alpha_{\text{electronic}}}{\alpha_{\text{electronic}} + \alpha_{\text{ionic}}} = 1 - \left(\frac{n^2 - 1}{n^2 + 2} \times \frac{\varepsilon_r + 2}{\varepsilon_r - 1}\right)$$

or, $$1 - \left(\frac{n^2 - 1}{n^2 + 2} \times \frac{\varepsilon_r + 2}{\varepsilon_r - 1}\right) = 1 - \left(\frac{(1.5)^2 - 1}{(1.5)^2 + 2} \times \frac{5.6 + 2}{5.6 - 1}\right) \left[\because n = 1.5 \text{ and } \varepsilon_x = 5.6\right]$$

$$= 0.514.$$

$\therefore$ percentage of contribution = **51.4%**.

## REVIEW QUESTIONS

1. Explain the following terms:
   (i) Polarization, (ii) Electric susceptibility, (iii) Atomic polarizability, (iv) Local field.
2. Derive the Claussius-Mossotti relation between polarizability and dielectric constant of a solid.
3. Write notes on:
   (i) Electronic polarizability
   (ii) Ionic polarizability
   (iii) Dipolar polarizability.
4. What do you mean by the following?
   (i) Ferroelectricity
   (ii) Piezoelectricity.
5. How does the total polarizability depend on frequency?

## PROBLEMS

1. Silicon has the dielectric constant 12 and the edge length of the conventional cubic cell of silicon lattice is 5.43 Å. Calculate the electronic polarizability of silicon atom.

   [**Ans.** $\alpha_{electronic} = 4.5 \times 10^{-40}$ Fm$^2$]
2. The optical refractive index and dielectric constant for water are 1.33 and 8.1 respectively. Determine the percentage of ionic polarizability. [**Ans.** 71%]

# 10
# *Short Biographies of Noted Solid State Physicists*

## 10.0 INTRODUCTION

In this chapter author is interested in writing short biographies of the noted solid state physicists who have contributed to the development of Solid State Physics.

## 10.1 HENDRIK ANTON LORENTZ

Dutch Physicist Hendrik Anton Lorentz was born on 18th July 1853 in Arnhem, Gelderland and died on 4th February 1928 at the age of 75. From 1866 to 1869 he attended the high school in Arnhem and in 1870 he passed the examination in classical languages. Lorentz studied Physics and Mathematics at the University of Leiden. In 1875 Lorentz earned a doctoral degree at the age of 22 years on "the theory of reflection and refraction of Light". In 1878 Lorentz at the age of 24 years was appointed professor of Theoretical Physics at the University of Leiden. Lorentz worked on the theory of electromagnetism. He also made contributions to mechanics, thermodynamics, kinetic theory, solid state physics and light. His most important contribution were in the area of electromagnetism, the electron theory and relativity. Lorentz made the theoretical interpretation of Zeeman effect. Lorentz's name is associated with Lorentz-Lorentz formula, Lorentz force, Lorentzian distribution, Lorentz transformation, Fitz Gerald-Lorentz contraction, time dilation. Lorentz was appointed the directror of research at Taylers Museum in Haarlem in 1912. He received Nobel Prize in Physics in 1902 for his work on influence of magnetism on the phenomena of atomic radiation. He was elected a Fellow of the Royal Society in 1905 and was honoured with Rumford Medal in 1908 and Copley Medal in 1918.

## 10.2 WILLIAM HENRY BRAGG

William Henry Bragg was born at Westward, Cumberland, England on July 2, 1862 and died on March 15, 1942 at the age of 80 years. He was educated at Market Harborough Grammer School and afterward at King William College. Elected a minor scholar of Trinity College, Cambridge in 1881. He was the third Wrangler in the Mathematics Tripos Part I in June 1884 and was placed in first class in Part II in following January, 1885. He studied Physics in Cavendish Laboratory during the first part of 1885 and at the end of that year he was appointed Professor of Mathematics and Physics in the University of Adelaide, South Australia. From 1909 to 1915 he was Cavendish Professor of Physics at Leeds and from 1915 to 1925 he was Quain professor of Physics at University College, London and Fullerian Professor of Chemistry in the Royal Institute.

In 1913-1914 Prof. W.H. Bragg and his son William Lawrence Bragg made analysis of crystal structure by means of X-rays. During the First World War Prof. W. H. Bragg worked on the detection and measurement of underwater sound in connection with the location of submarines. Prof. W.H. Bragg was elected Fellow of Royal Society in 1907, and in 1935 he was elected President of Royal Society. He was Knighted in 1920, the Order of Merit followed in 1931. He was Honorary doctorate of some sixteen Universities. He was awarded Rumford Medal in 1919 and Copley Medal in 1930. In 1915 Prof. W. H. Bragg was awarded Nobel Prize in Physics jointly with his son Sir William Lawrence Bragg for their work on the study of crystal structure by means of X-rays.

## 10.3 PIERRE ERNEST WEISS

French Physicist Pierre Ernest Weiss was born on March 25, 1865 at Mulhouse, France and died at the age of 75 on October 24, 1940 at Lyon, France. He is one of the Founders of Physics of magnetism. From 1883 to 1886 he studied Mechanical Engineering at ETH in Zurich and in 1887 he went to Ecole Normale Superiere in Paris. In 1886 he presented his thesis at the Faculte des Science in Paris. The theory was put forward on the magnetic properties of magnetite and iron-antimony alloys and he established the relation between magnetism and crystal symmetry. He was attached to Maitre de conference at Lyon. He became Professor and Director of Physics Laboratory at the ETH in Zurich in 1902. Here in 1906-1907 he formulated the molecular field hypothesis, magnetic domain theory. In 1907 he discovered the magnetic moment of atoms of ferromagnetic materials. In 1911 he discovered the existence of magneton. The Curie-Weiss law describes magnetic susceptibility of a ferromagnet. In 1918 Pierre Weiss went to Strassburg where he founded a Magnetic Laboratory and directed it for twenty years. Pierre Weiss became a member of Academic des Science in 1926.

## 10.4 ARNOLD SOMMERFELD

Arnold Sommerfeld was born on December 5, 1868 in Königsberg, Prussia [now Kaliningard, USSR] and died on 26th April, 1951 at the age of 83. Sommerfeld entered the Altstädti sches Gymnasium in Königsberg in 1875. He left the school after completing school examination and entered the University of Königsberg. In 1891 Sommerfeld was awarded his doctorate degree from Könisberg for his thesis "The arbitrary function in Mathematical Physics" under the supervision of Prof Lindemann. In 1893 he spent a year as an assistant in Mineralogical Institute. In 1894 he became Klein's assistant. In 1897 Sommerfeld became Professor of mathematics at the Mining Academy at Clausthal. In 1900 he became Professor of Mechanics at the Technische Hochschule in Aachen. In 1906 he become Professor of Theoretical Physics in Munich. Sommerfeld continued teaching until 1947 long after the usual retirement age of 65. Sommerfeld's impressive piece of work is on mathematical theory of diffraction, the work contains important theory of partial differential equations. He has contributed to the study of the propagation of electromagnetic waves in wires and the study of field produced by a moving electron. Sommerfeld along with Prof. Klein in 1909–1910 wrote a four volume text on the theory of gyroscopes. He worked on atomic spectra at Munich. He studied the hypothesis that X-rays were waves and proved this by using crystal as three-dimensional diffraction grating. In 1911 Sommerfeld proposed elliptical orbit of the atom in place of circular orbit propounded by Prof Neel Bohr. He introduced the magnetic quantum number in 1916, and in 1920 his work led to the discovery of electron spin. He used statistical mechanics to explain electronic properties of metals.

Sommerfeld won the Lorentz Gold Medal, the Planck Medal and Oersted Medal. He was elected to the Royal Society of London, the National Academy of Science in Washington, the Academis of Berlin, Munich, Vienna, Gottingen, Budapest, Uppsala and Madrid, Academia dei Lincei of Rome, the Academy of the USSR in Moscow and Indian Academy of Sciences. He received honourary degrees from many universities including Calcutta, Athens, Rostock and Aachen.

## 10.5 PAUL LANGEVIN

French Physicist Paul Langevin was born in Paris in 1872. He enrolled at an enginerring School where his Laboratory was supervised by Pierre Curie and his wife Marie Curie and Antsine Hernry Becquerrel. He received his PhD in 1902 under the guidance of Curies. He was professor of experimental Physics at the College de France from 1909. He is noted for his work on electron theory of magnetism and for his research on sound devices for submarine detection. Best work of Langevin in collaboration with Constantin Chilowski, is on the echolocation system, i.e., on sonar. Langevin concentrated his effort towards the study of Paramagnetism and Diamagnetism. Langevin died in 1946 at the age of 74.

## 10.6 ALBERT EINSTEIN

Prof. Albert Einstein was born in Ulm in Germany in 1879 and died at the age of 76 in 1955. He grew up in Munich. Einstein went to Switzerland in 1894 at the age of 16 to seek entry to the prestigious Swiss Federal Polytechnic (ETH) and was admitted to ETH in 1896 and he completed his ETH diploma in 1900. In ETH he had intimate friends, viz., Marcel Grossmann, Michelo Besso. He fell in love with a fellow student Mileva Maric and married with Mileva in 1902. He obtained a post at the Bern patent office as a patent clerk. In 1917 Einstein was appointed Director in the Institute of Physics at Berlin. In 1933 he joined the staff of the newly created Institute for advanced study in Princeton, New Jersey and he served the Institute until his death. In 1905 Einstein published three revolutionary research papers on (1) special theory of relativity (2) photoelectric effect (3) Brownian movement. Notable contribution of Einstein's work was on the Electrodynamics of moving bodies which was published in "Annal der Physik" in 1905 and is known as special theory of relativity. He extended and elaborated the work in 1915 to the general theory of relativity. Einstein is familiar with most people for his mathematical equation about the nature of energy, $E = mc^2$. Einstein in 1905 proposed the correct explanation of the photoelectric phenomenon by an extension of the quantum idea of Planck in which Einstein postulated that light could behave as lumps of energy or quanta and this is an extension of the quantum idea of Planck. This extension is known as Einstein's light quantum hypothesis. He argued that light can act as a discrete independent particle of energy in the same way as the particle of a gas. Einstein showed that light quanta as he called the particle of energy could help to explain phenomena being studied by experimental physicist; for example, he made clear how light ejects electrons from metals.

Einstein propsed the theory of the translation of Brownian movement which explains heat as the effect of the ceaseless motion of the atoms and he had created a powerful new tool for studying the movement of the atoms Einstein developed a theory about the relationship of mass and energy. The formula $E = mc^2$ is probably most famous outcome from Einstein's special theory of relativity. $E = mc^2$ means mass is just one form of energy and since speed of light ($3 \times 10^8$ m/sec) is very high a small amount of mass can be converted to huge amount of energy. The atom bomb project was taken by President Franklin D. Roosevelt after the letter of Einstein on 2.8.1939. In 1921 Einstein won the Nobel Prize in Physics for his discovery of the law of the photoelectric effect.

## 10.7 MAX VON LAUE

Max von Laue was born on october 9, 1879 in Pfaffenderf, Germany and died at the age of 80 on 24th April, 1960. He studied in the three cities Pasen, Berlin and Strassburg. In 1899 he studied mathematics, physics and chemistry in Strassburg and soon he moved to the University of Gottingen. He read a semester at the University of Munich, then he went to the University of Berlin. He obtained his doctorate degree at Berlin in 1903. In 1905 he joined the post of assistant to Max Planck at the Institute for Theoretical Physics at Berlin. In 1909 he joined as Privatdozen to the University of Munich. He became Professor of Physics at the University of Zurich. In 1919 he was appointed Professor of Physics at the University of Berlin and he held the post until 1943. He was the consultant to the Physikalish-Technische Reichsanstalt at Berlin-Charlottenberg from 1934. Von Laue was the second Director in the Institute for Physics at Berlin. He was acting Director of Max Planck Institute in 1946 and Titular Professor in the University. In 1951 he was elected Director of Fritz Haber Institute for Physical Chemistry.

Prof. von Laue received the Nobel Prize in Physics in the year 1914 for his work on diffraction of X-rays on crystals. He also contributed to the problem of superconductivity in 1932. He was awarded Ladenburg Medal, the Max Planck Medal, the Bimala Churn Law Gold Medal of Indian Association for the Cultivation of Science at Calcutta, India. He held honorary doctorate from the University of Berlin, Bonn, Stuttgart, Manchester and Chicago. He was a member of Russian Academy, Academy of Sciences of Berlin and the German Physical Society. In 1948 he became honorary President of International Union of Crystallographers. He was awarded knight title in 1953. He received Grand Cross with star for Federal Services and became an Officer of Legion of Honour of France in 1957.

## 10.8 WILLIAM LAWRENCE BRAGG

William Lawrence Bragg, son of William Henry Bragg was born in Adelaide, Australia on April 30, 1890 and died at the age of 81 on July 1, 1971. He received his early education at St. Peter's College in Adelaide and then took his degree in mathematics with first class honours in 1908. He entered Trinity College, Cambridge, England as an Allen Scholar and qualified with first class honours in the Natural Science Tripos in 1912. In 1914 he was appointed as Fellow and Lecturer in Natural Science at Trinity College. From 1915 to 1919 W. L. Bragg served as Technical Advisor on sound ranging to the Map Section, GHQ. From 1919 to 1937 he served as a Longworthy Professor of Physics at Manchester University. He was the Director of National Physical Laboratory in 1937–1938 and Cavendish Professor of Experimental Physics from 1938 to 1953. He was Chairman of the Frequency Advisory Committee from 1958 to 1960. Prof. W. L. Bragg published scientific papers on Crystal Structure, the Crystalline State, Electricity and Atomic Structure of Minerals. He was awarded Bernard Medal in 1914. He received Nobel Prize for Physics along with his father

Prof W. H. Bragg in 1915 for their work on the study of crystal structures by means of X-rays. He was elected Fellow of the Royal Society in 1921, knighted in 1941. He was honoured with the honorary DSc (Dublin, Leeds, Manchester, Lisbon, Paris, Brussels, Liege and Durham), honorary PhD (Coloque), honorary LLD (St. Andrew). He was honoured with many honorary fellowship and was an honorary or foreign member of American, French, Swedish, Chinese, Dutch and Belgian Scientific Academies. He was awarded the Hughes Medal of the Royal Society in 1931, Royal Medal of the same society in 1946 and Roebling Medal of the Mineral Society of America in 1948.

## 10.9 SATYENDRANATH BOSE

Indian scientist Prof. Satyendranath Bose was born on 1st January, 1894 and died on 4th February 1974 at the age of 80. Prof. Bose started his career in 1916 as a lecturer in Physics in Calcutta University, later he joined in Dacca University in 1921 as a lecturer in Physics. In 1924 he was promoted to reader in the Dacca University and continued teaching in the University until 1945. He then joined Calcutta University and taught there until 1956. Prof Bose was selected National Professor and held that post till death. Most important work of Prof. Bose is on Planck's quantum radiation law (1924). This work brought Prof. Bose closer to Prof. Albert Einstein who helped him to publish the work. This work enabled Prof. Bose to earn international recognition. As a result when he worked with internationally famous scientists like Prof. Louis de Broglie, Prof. Marie Curie and Prof. Albert Einstein, scientists referred to boson in the name of Prof. Bose acknowledging that particle obeys Bose statistics. Prof. Bose also published Planck's law and the hypothesis of light quanta. He published and worked on X-ray crystallography, unified field theories, the equation of state of real gases. He also worked in biochemistry.

Prof. Bose became a Fellow of Royal Society, London, England in 1958. In 1944 Prof. Bose was elected General President of the Indian Science Congress.

## 10.10 WOLFGANG ERNST PAULI

Prof. Wolfgang Ernst Pauli was born in Vienna on 25th April, 1900 and died on 15th December, 1958 at the age of 58. Wolfgang studied Mathematics and Physics at the Döblingen Gymnasium in Vienna and he graduated from the Gymnasium on July, 1918 with distinction. He then entered Ludwig-Maximilian University at Munich. Pauli received his doctorate degree in 1921 under the supervision of Prof. Sommerfeld for a thesis on quantum theory of ionised molecular hydrogen. In 1921 Pauli was appointed to Gottingen as Born's assistant. Pauli joined in Bohr's Institute in 1922 on the invitation of Prof. Bohr. He then joined University of Hamburg in 1923. In 1928 he was appointed professor of Theoretical Physics at the Federal Institute of Technology in Zurich, Switzerland. Dr. Pauli was the visiting professor at the University of Michigan in 1931 and at the Institute for Advance Study, Princeton in 1935–1936. In 1940 he was appointed to the chair of Theoretical Physics. In 1941 he was visiting professor at the University of Michigan and again in 1942 at the Purdue University. Pauli worked on the theory of relativity in 1921. Pauli proposed a quantum spin number for electrons. In 1925 he proposed his famous exclusion principle. In 1933 he predicted the existence of a particle of zero mass, later known as *neutrino* from 1934.

Pauli was awarded the Lorentz medal in Amsterdam on October, 1931. Pauli received Nobel Prize for his discovery of Quantum Exclusion Principle in 1945. He was elected a Fellow of the Royal Society of London in 1953. He was elected a member of the Swiss Physical Society, the American Physical Society and the American Association for the Advancement of Science.

## 10.11 WALTER HONSER BRATTAIN

American Scientist Walter Honser Brattain was born on February 10, 1902 at Amoy, China and died on October 13, 1987 in Beattle Wash, U.S. at the age of 85 years. He won Nobel Prize in 1956 in Physics for the work on invention and development of transistor. He worked in Bell Laboratory and retired in 1967.

## 10.12 FELIX BLOCH

Felix Bloch was born in Zurich, Switzerland on October in 1905 and died on 10th September, 1983 at the age of 78. From 1912 to 1918 he attended the public primary school and subsequently the Gymnasium in Canton in Zurich up to 1924. He then entered in Federal Institue of Technology in Zurich to study engineering course. After one year he changed his subject to Physics in the same institute. He received his degree of Doctor of Philosophy in 1928 in quantum mechanics of electrons in crystals and developed the theory of metallic conduction. In 1933 he left Germany and joined at Stanford University, U.S.A. In 1939 he collaborated with L.W. Alvaretz at the Barkeley Cyclotron Laboratory in which the magnetic moment of neutron was determined. In 1954 Bloch joined as the first Director General of CERN in Geneva. In 1961 he received an endowed Chair by his appointment as Max Stein Professor of Physics at Stanford University. He was awarded the Nobel prize for physics in 1952.

## 10.13 JOHN BARDEEN

American Physicist John Bardeen was born in Medison, Wisconsin on May 23, 1908 and died at the age of 82 years on January 30, 1991. Bardeen studied both Mathematics and Physics in his undergraduate course. He submitted his thesis in Solid State Physics and received PhD degree in 1936 in Mathematical Physics from Princeton University.

He was Junior fellow at Harvard University in 1935. In 1938 he joined as Assistant Professor at the University of Minnesota. In 1945 he began work at Bell Laboratory as a member of a Solid State Physics Group. Bardeen and Brattain succeeded in creating point-contact transistor in 1947. By 1951 Bardeen joined the University of Illinois at Urbanachampaign. He was Professor of Electrical Engineering and Physics at Illinois from 1951 to 1975.

In 1956 John Bardeen was awarded Nobel Prize in Physics with William Shockley and Walter Brattain for their invention and development of point-contact transistor.

In 1972 John Bardeen shared the Nobel Prize in Physics with Neil Cooper of Brown University and John Robert Schrieffer of the University of Pennsylvania for their jointly developed theory of superconductivity usually known as BCS theory. Bardeen was the first person to win Nobel Prize twice in Physics.

## 10.14 WILLIAM BRADFORD SHOCKLEY

American Physicist William Bradford Schockley was born in February 13, 1910 in London and died on 12th August 1989 at the age of 79. He got his Bachelor of Science degree from the California Institute of Technology in 1932. Shockley was awarded PhD from Massachusettes Institute of Technology in 1936. The title of his doctoral thesis was "Electronic bonds in Sodium Chloride". He joined in research project of Bell Laboratory in New Jersey. After the World War II he joined Bell Lab's Solid State Physics Group. Shockley developed sandwich transistor, i.e., junction transistor.

In 1951 he was elected a member of the National Academy of Science (NAS). In 1953 he was chosen as the recepient of the prestigious Comstock Prize for Physics by the NAS and was the recipient of many other awards and honours. Shockley was a co-recipient of the Nobel Prize in Physics in 1956 along with Bardeen and Brattain for their invention and development of transistor.

# Appendix

## A. BOSE-EINSTEIN STATISTICS

The Bose-Einstein statistics considers a system of $N$ number of identical and indistinguishable particles which do not obey Pauli's exclusion principle. The particles that obey the Bose-Einstein statistics are called *bosons*.

Consider a system of $N$ number of identical and indistinguishable particles among $g_i$ quantum states of which the total energy of the system is $E$. Temperature and volume of the system are $T$ and $V$ respectively. Let there be $K$ number of discrete energy levels associated with bosons.

The situation is displayed as follows:

Energy levels $\varepsilon_1, \varepsilon_2, ..., \varepsilon_K$

Degeneracies $g_1, g_2, ..., g_K$

Number of bosons $n_1, n_2, ..., n_K$

The system obeys the following constraints:

$$n_1 + n_2 + n_3 + \cdots + \cdots + n_K = \sum_{i=1}^{K} n_i = N \tag{1}$$

and

$$n_1\varepsilon_1 + n_2\varepsilon_2 + n_3\varepsilon_3 + \cdots + \cdots + n_K\varepsilon_K = \sum_{i=1}^{K} n_i\varepsilon_i = E\,. \tag{2}$$

The number of ways in which $N$ particles can be distributed in $g_i$ states is given by

$$W_i = \frac{(n_i + g_i - 1)!}{n_i!\,(g_i - 1)!}\,.$$

So the number of microstates available in the whole system is

$$W = \prod_{i=1}^{K} \frac{(n_i + g_i - 1)}{n_i!(g_i - 1)!}$$

$$W_i = \frac{(n_1 + g_1 - 1)}{n_i!(g_1 - 1)!} \times \frac{(n_2 + g_2 - 1)!}{n_2!(g_2 - 1)!} \times \cdots \times \frac{(n_K + g_K - 1)!}{n_K!(g_K - 1)!}$$

$$= \prod_{i=1}^{K} \frac{(n_i + g_i - 1)!}{n_i!(g_i - 1)!},$$

where $\Pi$ denotes product.

Now applying Sterling's theorem for large $n_i$ and $g_i$ and neglecting 1 in comparison to $n_i$ and $g_i$, we get

$$\ln W = \sum_{i=1}^{K} \left[(n_i + g_i) \ln (n_i + g_i) - (n_i + g_i) - n_i \ln n_i + n_i - g_i \ln g_i + g_i\right]$$

$$= \sum_{i=1}^{K} \left[(n_i + g_i) \ln (n_i + g_i) - n_i \ln n_i + \delta ni - g_i \ln g_i\right]$$

Now maximizing the above equation with respect to $n_i$, we get

$$\delta \ln W = \sum_{i=1}^{K} \left[\delta n_i \ln (n_i + g_i) + \frac{(n_i + g_i)}{(n_i + g_i)} \delta n_i - n_i \frac{\delta n_i}{n_i} - \delta n_i \ln n_i\right] = 0$$

$$= \sum_{i=1}^{K} \ln \frac{n_i + g_i}{n_i} \delta n_i = 0. \tag{3}$$

From equations (1) and (2), we get

$$\sum_{i=1}^{K} \delta n_i = 0 \tag{4}$$

and

$$\sum_{i=1}^{K} \varepsilon_i \, \delta n_i = 0. \tag{5}$$

From equations (3), (4) and (5) applying the Lagrange method of undetermined multiplier, we get

$$\sum_{i=1}^{K} \left[\ln \frac{n_i + g_i}{n_i} - \alpha - \beta \varepsilon_i\right] \delta n_i = 0 \tag{6}$$

where $\alpha$ and $\beta$ are two constants to be determined. As the variations of $\delta n_i$ are independent of each other the equation (6) can be true, only if

$$\ln \frac{n_i + g_i}{n_i} - \alpha - \beta \, \varepsilon_i = 0.$$

or,

$$\frac{n_i + g_i}{n_i} = e^{\alpha + \beta \varepsilon_i}$$

or, $$1+\frac{g_i}{n_i}=e^{\alpha+\beta\varepsilon_i}$$

or, $$n_i=\frac{g_i}{e^{\alpha+\beta\varepsilon_i}-1}. \tag{7}$$

The equation (7) gives the most probable distribution of bosons among various energy levels.

The quantity $\beta$ can be calculated from thermodynamic consideration as

$$\beta=\frac{1}{KT},$$

where $K$ is Boltzmann constant.

Constant $\alpha$ is given as

$$\alpha=-\frac{\mu}{KT},$$

where $\mu$ is the chemical potential.

Therefore, the equation (7) becomes

$$n_i=\frac{g_i}{e^{(\varepsilon_i-\mu)/KT}-1}. \tag{8}$$

The distribution of energy in Bose-Einstein statistics is given by

$$n(\varepsilon)\,dE=\frac{g(\varepsilon)d\varepsilon}{e^{(\varepsilon_i-\mu)/KT}-1}. \tag{9}$$

It is readily seen from equation (9) at absolute zero that the particle tends to occupy the lowest energy state.

## B. FERMI-DIRAC STATISTICS

Fermi-Dirac statistics assumes that the particles are identical and indistinguishable and these particles obey Pauli's exclusion principle which states that no two electrons in an atom can have the same four quantum numbers. In other words, it excludes two protons from occupying the same quantum state, and two neutrons from having the same quantum numbers. The subatomic particles which obey Fermi-Dirac statistics are called *Fermions*.

Consider a system of $N$ number of identical and indistinguishable fermions. Let the total energy of the system be $E$, volume $V$ and temperature $T$ respectively. Assume, there are $K$ number of discrete energy levels associated with the fermions. The situation of distribution is as follows:

| | |
|---|---|
| Energy levels | $\varepsilon_1, \varepsilon_2, \varepsilon_3, ..., \varepsilon_K$ |
| Degeneracies | $g_1, g_2, g_3, ..., g_K$ |
| Occupation number | $n_1, n_2, n_3, ..., n_K$ |

The basic postulates in Fermi-Dirac statistics are the following:

1. Fermions are indistinguishable and identical from each other, hence the rearrangement of fermions among themselves does not lead to new microstates.
2. Due to Pauli's exclusion principle each sub-level may contain zero or one particle, i.e., $g_i$ must be greater than or equal to $n_i$.

3. The total number of paritcles are fixed and the total energies of all the particles are also fixed, i.e.,

$$\sum_{i=1}^{K} n_i = N \tag{10}$$

$$\sum_{i=1}^{K} n_i \varepsilon_i = E . \tag{11}$$

The number of ways in which $n_i$ particles are arranged in $g_i$ quantum states is given by

$$W_i = {}^{g_i}C_{n_i} = \frac{g_i\,!}{n_i\,!\,(g_i - n_i)!} .$$

Hence the total number of microstates available to the system

$$W = \prod_{i=1}^{K} \frac{g_i\,!}{n_i\,!\,(g_i - n_i)!} , \tag{12}$$

where $\Pi$ denotes the product. Applying Sterling's theorem to the equation (12), we get

$$\ln W = \sum_{i=1}^{K} \left[ g_i \ln g_i - g_i - n_i \ln n_i + n_i - (g_i - n_i) \ln (g_i - n_i) + g_i - n_i \right]$$

$$= \sum_{i=1}^{K} \left[ g_i \ln g_i - n_i \ln n_i - (g_i - n_i) \ln (g_i - n_i) \right] . \tag{13}$$

Differentiating equation (13) with respect to $n_i$ , we have

$$\delta \ln W = \sum_{i=1}^{K} \left[ -\delta n_i \ln n_i - \frac{n_i}{n_i} \delta n_i + \delta n_i \ln (g_i - n_i) - \frac{(g_i - n_i)}{(g_i - n_i)} (-\delta n_i) \right]$$

$$= \sum_{i=1}^{K} \left[ \ln \frac{(g_i - n_i)}{n_i} \delta n_i \right] . \tag{14}$$

By the condition of maximum probability the equation (14) gives

$$\delta \ln W = \sum_{i=1}^{K} \left[ \ln \frac{(g_i - n_i)}{n_i} \delta n_i \right] = 0 . \tag{15}$$

Again, from equations (10) and (11), we get

$$\sum_{i=1}^{K} \delta n_i = 0 \tag{16}$$

$$\sum_{i=1}^{K} \varepsilon_i \, \delta n_i = 0 . \tag{17}$$

From equations (15), (16) and (17) and using Lagrange's method of undetermined multipliers, we have

$$\sum_{i=1}^{K}\left[\ln\left(\frac{g_i - n_i}{n_i}\right) - \alpha - \beta\varepsilon_i\right]\delta n_i = 0 .$$

Since $\delta n_i$ can be treated as arbitrary, we get

$$\ln\left(\frac{g_i - n_i}{n_i}\right) - \alpha - \beta\varepsilon_i = 0.$$

or,
$$\frac{g_i}{n_i} - 1 = e^{\alpha+\beta\varepsilon_i}$$

or,
$$n_i = \frac{g_i}{e^{\alpha+\beta\varepsilon_i} + 1}. \tag{18}$$

The equation (18) is the most probable distribution of fermions.

Now the constant $\beta$ is determined thermodynamically as

$$\beta = \frac{1}{KT}$$

and usually $\alpha$ is written as $\alpha = -\frac{\varepsilon_F}{KT}$,

where $\varepsilon_F$ is known as Fermi energy.

Hence the *Fermi-Dirac distribution* can be written as

$$n_i = \frac{g_i}{e^{(E_i-\varepsilon_F)/KT} + 1}.$$

For energy interval $\varepsilon$ and $\varepsilon + d\varepsilon$ and continuous distribution of energies the Fermi-Dirac distribution is given by

$$n(E)\,dE = \frac{g(E)\,dE}{e^{(E_i-\varepsilon_F)/KT} + 1}.$$

## C. TABLE OF PHYSICAL CONSTANTS AND ENERGY CONVERSION

### I. Table of Physical Constants

| | |
|---|---|
| Avogadro number | $N = 6.023 \times 10^{26}$ kg-mole$^{-1}$ |
| Boltzmann's constant | $k = 1.381 \times 10^{-23}$ joule per degree kelvin |
| Classical radius of the electron | $r_e = 2.8 \times 10^{-15}$ m |
| Compton wavelength of the electron | $\lambda_e = \frac{h}{m_0 c} = 2.426 \times 10^{-12}$ m |
| Electric permeability of the vacuum | $\varepsilon_0 = 8.854 \times 10^{-12}$ A sec/Vm |
| Stefan's radiation constant | $\sigma = 5.67 \times 10^{-8}$ W/m$^2$ K$^4$ |

| | |
|---|---|
| Magnetic permeability of free space | $\mu_0 = 4\pi \times 10^{-7}$ Hm$^{-1}$ |
| | $= 1.257 \times 10^{-6}$ Hm$^{-1}$ |
| Electric elementary charge | $e = 1.602 \times 10^{-19}$ C |
| Rest mass of electron | $m_e = 9.11 \times 10^{-31}$ kg |
| Rest mass of neutron | $m_n = 1.6749 \times 10^{-27}$ kg |
| Rest mass of proton | $m_p = 1.6726 \times 10^{-27}$ kg |
| Speed of light in vacuum | $c = 2.998 \times 10^8$ m/sec |
| Specific charge of the electron | $\dfrac{e}{m_e} = 1.759 \times 10^{11}$ C/kg |
| Planck's constant | $h = 6.626 \times 10^{-34}$ Js |
| | $\hbar = \dfrac{h}{2\pi} = 1.056 \times 10^{-34}$ Js |
| Atomic mass unit (12 units for $m_u$ the $^{12}$C) | $1.6605 \times 10^{-27}$ kg = 1 u |
| Gas constant | $R_0 = 8.314 \times 10^3$ J mol$^{-1}$ K$^{-1}$ |
| Rydberg constant for infinite nuclear mass | $R_\infty = 1.0974 \times 10^7$ m$^{-1}$ |
| Bohr radius for infinite nuclear mass | $a_0 = 5.292 \times 10^{-11}$ m |
| Bohr magneton | $\mu_B = \dfrac{e\hbar}{2m_e} = 0.9274 \times 10^{-23}$ A-mt |
| 1 ångström Unit (Å) | $= 10^{-10}$ m |
| 1 eV | $= 1.6 \times 10^{-19}$ joule |

## II. Energy Conversion

| | |
|---|---|
| 1 J | = 1 Ws |
| 1 eV | $= 1.6 \times 10^{-19}$ J |
| 1 cal | = 4.187 J |
| 1 Pa | = 1 Nm$^{-2}$ = $10^{-5}$ bar |
| 1 T | = 1 Wb m$^{-2}$ |
| 1 torr | = 133.3 Pa |

## D. DATA OF THE STANDARD METAL

| | |
|---|---|
| Fermi energy | 5.59 eV |
| Temperature of degeneracy | 25900 °K |
| Radius of the Fermi sphere | $1.21 \times 10^6$ m$^{-1}$ |
| Specific heat of electron per unit volume | .63 TJ/m$^3$ °K |
| Fermi velocity | $1.40 \times 10^6$ m/s |
| Cyclotron frequency of electrons in a magnetic field +1 | $2.21 \times 10^8$ H$\left[\dfrac{A}{m}\right]$ |
| Classical Cyclotron radius for electrons at the Fermi energy | 0.633 m |

# E. PERIODIC TABLE

| 1 | 2 | 3 | 4 | 5 | 6 | 7 | 8 | 9 | 10 | 11 | 12 | 13 | 14 | 15 | 16 | 17 | 18 |
|---|---|---|---|---|---|---|---|---|---|---|---|---|---|---|---|---|---|
| 1<br>H<br>1.0079 | | | | | | | | | | | | | | | | | 2<br>He<br>4.0026 |
| 3<br>Li<br>6.941 | 4<br>Be<br>9.0122 | | | | | | | | | | | 5<br>B<br>10.811 | 6<br>C<br>12.011 | 7<br>N<br>14.007 | 8<br>O<br>15.999 | 9<br>F<br>18.998 | 10<br>Ne<br>20.180 |
| 11<br>Na<br>22.990 | 12<br>Mg<br>2.4305 | | | | | | | | | | | 13<br>Al<br>26.982 | 14<br>Si<br>28.086 | 15<br>P<br>30.974 | 16<br>S<br>32.065 | 17<br>Cl<br>35.453 | 18<br>Ar<br>39.948 |
| 19<br>K<br>39.098 | 20<br>Ca<br>40.078 | 21<br>Sc<br>44.956 | 22<br>Ti<br>47.867 | 23<br>V<br>50.942 | 24<br>Cr<br>51.996 | 25<br>Mn<br>54.938 | 26<br>Fe<br>55.845 | 27<br>Co<br>58.933 | 28<br>Ni<br>58.693 | 29<br>Cu<br>63.546 | 30<br>Zn<br>65.38 | 31<br>Ga<br>69.723 | 32<br>Ge<br>72.64 | 33<br>As<br>74.922 | 34<br>Se<br>78.96 | 35<br>Br<br>79.904 | 36<br>Kr<br>83.798 |
| 37<br>Rb<br>85.468 | 38<br>Sr<br>87.62 | 39<br>Y<br>88.906 | 40<br>Zr<br>91.224 | 41<br>Nb<br>93.906 | 42<br>Mo<br>95.96 | 43<br>Tc<br>(98) | 44<br>Ru<br>101.07 | 45<br>Rh<br>102.91 | 46<br>Pd<br>106.42 | 47<br>Ag<br>107.87 | 48<br>Cd<br>112.41 | 49<br>In<br>114.82 | 50<br>Sn<br>118.71 | 51<br>Sb<br>121.76 | 52<br>Te<br>127.60 | 53<br>I<br>126.90 | 54<br>Xe<br>131.29 |
| 55<br>Cs<br>132.91 | 56<br>Ba<br>137.33 | 57-71<br>* | 72<br>Hf<br>178.49 | 73<br>Ta<br>180.95 | 74<br>W<br>183.84 | 75<br>Re<br>186.21 | 76<br>Os<br>190.23 | 77<br>Ir<br>192.22 | 78<br>Pt<br>195.08 | 79<br>Au<br>196.97 | 80<br>Hg<br>200.59 | 81<br>Tl<br>204.38 | 82<br>Pb<br>207.2 | 83<br>Bi<br>208.98 | 84<br>Po<br>(209) | 85<br>At<br>(210) | 86<br>Rn<br>(222) |
| 87<br>Fr<br>(22.3) | 88<br>Ra<br>(22.6 | 89-103<br># | 104<br>Rf<br>(261) | 105<br>Db<br>(262) | 106<br>Sg<br>(266) | 107<br>Eh<br>(264) | 108<br>Hs<br>(270) | 109<br>Mt<br>(268) | 110<br>Ds<br>(281) | 111<br>Rg<br>(272) | 112<br>Uub<br>(285) | 113<br>Uut<br>(284) | 114<br>Uuq<br>(289) | 115<br>Uup<br>(288) | 116<br>Uuh<br>(291) | | 117<br>Uuo<br>(294) |

| | | | | | | | | | | | | | | | |
|---|---|---|---|---|---|---|---|---|---|---|---|---|---|---|---|
| * Lanthanide series | 57<br>La<br>138.91 | 58<br>Ce<br>140.12 | 59<br>Pr<br>140.91 | 60<br>Nd<br>144.24 | 61<br>Pm<br>(145) | 62<br>Sm<br>150.36 | 63<br>Eu<br>151.96 | 64<br>Gd<br>157.25 | 65<br>Tb<br>158.93 | 66<br>Dy<br>162.50 | 67<br>Ho<br>164.93 | 68<br>Er<br>167.26 | 69<br>Tm<br>168.93 | 70<br>Yb<br>173.05 | 71<br>Lu<br>174.97 |
| # Actinide series | 89<br>Ac<br>(227) | 90<br>Th<br>232.04 | 91<br>Pa<br>231.04 | 92<br>U<br>238.03 | 93<br>Np<br>(237) | 94<br>Pu<br>(244) | 95<br>Am<br>(243) | 96<br>Cm<br>(247) | 97<br>Bk<br>(249) | 98<br>Cf<br>(251) | 99<br>Es<br>(252) | 100<br>Fm<br>(257) | 101<br>Md<br>(258) | 102<br>No<br>(259) | 103<br>Lr<br>(262) |

# *Bibliography*


Adrianus, J. Dekker, *Solid State Physics*, Macmillan India Ltd. (Delhi, 1998).

Alexander, O.E. Animaln, *Intermediate Quantum Theory of Crystalline Solids*, Prentice-Hall of India Pvt. Ltd. (New Delhi, 1978).

Azaraff, L.V., *Introduction to Solid State Physics*, Tata McGraw-Hill Publishing Co., (New Delhi, 2003).

Charles Kittel; *Introduction to Solid State Physics*, 3rd Ed, Wiley Eastern Pvt. Ltd., (New Delhi, 1971).

Chattopadhyay, D. and P.C. Rakshit, *Quantum Mechanics, Statistical Mechanics and Solid State Physics*, S. Chand & Company Ltd. (New Delhi, 1989).

George Busch and Horst Schade, *Lectures on Solid State Physics*, Vol. 79, Pergamon Press Oxford, (New York, 1976).

Madelung, O., *Introduction to Solid State Theory*, Springer-Verlag, Berlin, 2nd reprint, (1981).

Pillai, S.O., *Solid State Electronic Engineering Materials*, Wiley Eastern Ltd. (New Delhi, 1992).

Puri, R.K., Babbar, V.K., *Solid State Physics*, S. Chand & Company Ltd. (New Delhi, 2006).

Saxena, B.S., Gupta, R.C., Saxena, P.N., *Fundamentals of Solid State Physics*, Pragati Prakashan, (Meerut, 1993).

Ziman, J.M., *Principles of the Theory of Solids*, (Cambridge, 1972).

# Author Index


Bragg, W.H. 142
Bragg, W.L. 144
Boltzmann, L. 44
Bloch, F. 59, 146
Bohr, N. 84
Bardeen, J. 146
Brattain, W.H. 146
Bravais, A. 4
Bose, S.N. 145

Claussius, R. 131
Cooper, I. 123
Curie, M. 84
Curie, P. 143

de Broglie, L. 51
Debye, P. 27
Dirac, P.A.M. 46
Drüde, P. 44

Einstein, A. 143

Fermi, E. 149
Fity Gerald, F. 141
Franz, E.N. 45
Frenkel, A. 152
Friedrich, W. 24

Giaever, I. 123

Heitler, W. 58
Hall, H.E. 73
Hundy, E. 87
Hand, F 78

Jones, L 40
Josephson, B. 124

Kamerlingh, Onnes 115
Knipping, P. 24
Kronig, R. 63

Laue, M. 24
Langevin, P. 82
London, F. 119
London, H. 40
Lorentz, H.A. 44

Madelung, H. 22
Maxwell, J.C. 44
Meissner, W. 116
Miller, R.C. 14
Mossetti, O.F. 131
Mosley, H 22

Neel, L. 92

Pauli, W.E. 145
Penney, G. 63
Röntgen , W.K. 19

Scherrer, P. 27
Schrieffer, J.R. 123
Schrödinger, E. 59
Shockley, W.B. 146
Schottkey, H. 113
Sommerfeld, A. 46

Tuyn, N. 116

van der Waals J.D. 40

Weiss, P.E. 88
Wiedemann, G.H. 45

# *Subject Index*


Acceptor level 99
Allowed and forbidden band 67
Amorphous solid 1
Antiferromagnetism 91
Asymptotic Curie temperature 92
Atomic packing factor 7
Atomic polarizability 130
Atomic radius 7

Bloch 59
Bloch theorem 59
Bloch theorem in three dimensions 62
Bohr magneton 84
Boltzmann constant 82
Bragg's equation 24
Bragg's law 25
Brillouin zone 58
Burgers vector 113, 114

Carrier concentration 101
Carrier mobility 100
Claussius-Mossetti relation 131
Cohesive Energy 41
Conductivity 73
Conduction band 102
Cooper pair 123
Coordination number 7
Coulomb interaction potential energy 41
Covalent bonds 42
Curie temperature (Ferromagnetic) 91
Crystalline solid 1
Cryoelectronics 125
Crystal systems 4
Cubic 4
Curie temperature 91

Density of States 53
Diamagnetism 79
Dielectric 129
Dipolar polarizability 134
Dislocation 113
Dislocation density 114
Donor level 99
Doping 98
Drift velocity 73

Edge Dislocation 113
Effective mass of an electron 70
Effective number of Bohr magneton 86
Electrical conductivity 45
Electronic polarizability 133, 135
Electric polarization 129
Energy band gap 99
Extrinsic semiconductor 98

Fermi energy 49
Fermi levels 49, 53

Fermi-Dirac distribution 151
Ferrimagnetism 88
Ferroelectricity 126
Ferroelectric crystal 138
Free electron model 44
Free electron theory 44, 46
Frenkel defect 113

Germanium 96
Grain boundary defect 114
Gyromagnetic ratio 80

Hall coefficient 74
Hall effect 73
Hall voltage 74
Heitler-London valence band 58
Hexagonal 4
Hole 97
Hund's rule 87
Hydrogen bounded crystal 43

Impurity semiconductor 98
Insulator 78, 129
Intrinsic carrier concentration 98
Intrinsic semiconductor 98
Ionic crystals 41
Ionic polarizability 134

Josephson effect 124

Kronig-Penny model 63

Lande's *g* factor 84
Langevin function 87
Langevin's theory of paramagnetism 82
Langevin's theory of diamagnetism 79
Lattice 2
Lattice constant 13
Lattice defect 113
Laue's equation 24
Leonard-Jones potential 40
Local field 131
London equation 119
Lorentz relation 131

Madelung constant 42
Madelung energy 42
Magnetic field 78
Magnetization 78
Magnetic induction 78
Magnetic permeability 78
Magnetic moment 80
Magnetic susceptibility per unit volume 78
Metal 73
Metallic crystals 42
Miller indices 14
Monoclinic 5
Molar susceptibility 78
Molecular crystals 40
Molecular constant 89

*n*-type or Donor type semiconductor 99
Neel temperature 92
NMR spectrometer 125
Nuclear magnetic moment 88
Nuclear Paramagnetism 88

Orbital angular momentum 80
Oriental polarizability 134
Orthorhombic crystal system 5

*p*-type or acceptor type semiconductor 99
Paramagnetism 82
Paracrystalline 39
Paramagnetic susceptibility 84
Pauli's exclusion principle 40
Periodic table 99
Permittivity 131
Planar defects 114
Piezoelectricity 138
Point defects 113
Polarization 132
Primitive cell 3

Quantum number 89
Quantum theory of free electrons 46
Quantum theory of paramagnetism 84

Reciprocal lattice 18

Schrödinger equation 47
Screw Dislocation 113
Semiconductor 98
Semiconductor diode 98
Sommerfeld model of free electron 46
Spin quantum number 49
Spontaneons magnetization 88

Specific heat 118
Stacking fault 114
Susceptibility 131
Superconductivity 115
Superconducting quantum interference device (SQUID) 125

Temperature coefficient of resistance 96
Tesla 79
Triclinic 4
Trigonal 4
Translation vector 2
Transistor 96
Trivalent 99
Tunnelling 124

Unit cell 2

van der Waals bonding 41

Weber 79
Weiss field constant 89
Weiss theory of Ferromagnetism 88
Wiedemann-Franz law 45

*Some exciting gifts are up for grabs with this book!!*

# INTRODUCTION TO CRYSTALLOGRAPHY AND SOLID STATE PHYSICS

Debadideb Bhattacharya

FREE Gifts Rs. 1000/-

## DEAR READERS

By now you must have gone through this book in the entirety. So we invite your assessment on this title with absolute impartiality and candid observations. After all, we publishers blindly depend on your benchmarking. Friends, please fill up the open spaces below with CAPITAL LETTERS. You may add some extra sheets of paper if the space provided is not sufficient for you.

i. What are the important topics of your syllabus which are not adequately covered in this book?

ii. In which chapters and on what topics the treatment of the subject-matter by the author is not satisfactory and why?

iii. Have you come across the typos, factual inaccuracies, inconsistencies, and other forms of mistakes in the book? If yes, then please specify the mistakes and page numbers.

iv. Name top three books on the same subject, which, in your consideration, are better than this book. Could you provide some reasons?

Your 1st Preference: ..................................................

Your 2nd Preference: ..................................................

Your 3rd Preference: ..................................................

v. Any suggestion to improve the quality of illustrations, diagrams, etc.?

vi. How did you like the cover design? Any suggestion on colours?

vii. Suggestions for further improvement of production-standard:

viii. Who recommended you this book:

☐ Teacher ☐ Friend ☐ Parents ☐ Bookseller ☐ Librarian

ix. Name of the person who has recommended:

Designation: ..................................................

Address: ..................................................

Phone No.: ..................................................

e-mail: ..................................................

x. Name and address of the bookseller from whom you have purchased this book:

..........................................................................................................

..........................................................................................................

..........................................................................................................

xi. Name and address of the library where you would like to find the book:

..........................................................................................................

..........................................................................................................

xii. Name and address of your institution (Please mention the University or Board):

..........................................................................................................

..........................................................................................................

xiii. Your name, e-mail, and complete postal address:

..........................................................................................................

..........................................................................................................

..........................................................................................................

xiv. A few other titles published by us that you would like to read:

..........................................................................................................

..........................................................................................................

The best assessment will be rewarded half-yearly. The award will be in the form of our publications, as decided by the Editorial and Marketing Operations Division, amounting to Rs. 1000/- (Rupees one thousand) only.

..........................................................................................................

Please mail the filled up coupon at your earliest to :

*Editorial & Marketing Operations Division*


**NEW CENTRAL BOOK AGENCY (P) Ltd**

**8/1 Chintamoni Das Lane, Kolkata 700 009**

**West Bengal, INDIA**

***Phone + Fax:*** 033 + 2241 8411, 2241 8573, 2241 9094

***e-mail:*** ncbapvtltd@eth.net, ncba@cal2.vsnl.net.in

***Website:*** www.centralgroup.com